TEUBNER-TEXTE zur Informatik   Band 5

R. Zhao

Handsketch-Based Diagram Editing

TEUBNER-TEXTE zur Informatik

Herausgegeben von
Prof. Dr. Johannes Buchmann, Saarbrücken
Prof. Dr. Udo Lipeck, Hannover
Prof. Dr. Franz J. Rammig, Paderborn
Prof. Dr. Gerd Wechsung, Jena

Als relativ junge Wissenschaft lebt die Informatik ganz wesentlich von aktuellen Bei-
trägen. Viele Ideen und Konzepte werden in Originalarbeiten, Vorlesungsskripten und
Konferenzberichten behandelt und sind damit nur einem eingeschränkten Leserkreis
zugänglich. Lehrbücher stehen zwar zur Verfügung, können aber wegen der schnellen
Entwicklung der Wissenschaft oft nicht den neuesten Stand wiedergeben.

Die Reihe „TEUBNER-TEXTE zur Informatik" soll ein Forum für Einzel- und Sammel-
beiträge zu aktuellen Themen aus dem gesamten Bereich der Informatik sein. Gedacht ist
dabei insbesondere an herausragende Dissertationen und Habilitationsschriften, spezielle
Vorlesungsskripten sowie wissenschaftlich aufbereitete Abschlußberichte bedeutender
Forschungsprojekte. Auf eine verständliche Darstellung der theoretischen Fundierung und
der Perspektiven für Anwendungen wird besonderer Wert gelegt. Das Programm der
Reihe reicht von klassischen Themen aus neuen Blickwinkeln bis hin zur Beschreibung
neuartiger, noch nicht etablierter Verfahrensansätze. Dabei werden bewußt eine gewisse
Vorläufigkeit und Unvollständigkeit der Stoffauswahl und Darstellung in Kauf genommen,
weil so die Lebendigkeit und Originalität von Vorlesungen und Forschungsseminaren bei-
behalten und weitergehende Studien angeregt und erleichtert werden können.

TEUBNER-TEXTE erscheinen in deutscher oder englischer Sprache.

Handsketch-Based Diagram Editing

Von Rui Zhao

Universität-Gesamthochschule-Paderborn

 Springer Fachmedien
Wiesbaden GmbH 1993

Dr. rer. nat. Rui Zhao

Rui Zhao is born in 1962 at Shandong, China. He studied computer science and electrical engineering at University of Dortmund from 1982 to 1988. Since 1988 he is a computer scientist at Cadlab, a joint venture University of Paderborn and Siemens Nixdorf Informationssysteme AG. He received his Dr. rer. nat. in 1992 from the University of Paderborn. He is interested in user interface technology, computer-aided design, graphical editor, and pen-based computers.

Dissertation an der Universität-Gesamthochschule-Paderborn im Fachbereich Mathematik/Informatik

Die Deutsche Bibliothek – CIP-Einheitsaufnahme

Zhao, Rui:
Handsketch-based diagram editing / von Rui Zhao. –
Stuttgart ; Leipzig : Teubner, 1993
  (Teubner-Texte zur Informatik ; Bd. 5)
  Zugl.: Paderborn, Univ., Diss., 1992
  ISBN 978-3-322-95369-8   ISBN 978-3-322-95368-1 (eBook)
  DOI 10.1007/978-3-322-95368-1

NE: GT

Umschlaggestaltung: E. Kretschmer, Leipzig

Preface

This thesis concerns concepts and techniques of handsketch-based diagram editors. Diagram editing is an attractive application of gestural interfaces and pen-based computers which promise a new input paradigm where users communicate with computers in diagram languages by using gestures. Though recent advances of pen-based computer technology and pattern recognition methodology, developing gesture-based diagram editors is difficult. The key problem is the on-line gesture recognition which can be classified into two levels: one considers to recognize on-line sketched gestures formed by x-y coordinates into symbols and the other transforms these symbols into editing commands.

In the thesis, I discuss the key ideas of the incremental and the cooperative recognition, the gesture specification and the structure recognition, as well as decoupling the recognition interface from the command interface. For reducing the development efforts of handsketch-based diagram editors, an editor framework and two experimental applications are designed, implemented, and evaluated. The results indicate that the implementation efforts of such editors is drastically reduced by using the editor framework Handi. The Handi-based editors follows the so-called WYDIWYG input principle, they are easy to use and appropriate for conceptual sketching.

This dissertation has been carried out during my research activities in Cadlab, a cooperation University of Paderborn and Siemens Nixdorf Informationssysteme AG. While a dissertation is an individual effort, its magnitude and duration ensure that many others have assisted in its production.

There is much for which I have to thank my two advisors, Franz Rammig and Gerd Szwillus. Franz was the first to suggest me that pen-based user interface might be an interesting problem. I thank him to inspire me and to show me to do this research. I was lucky that Gerd came to University of Paderborn while I was beginning with this research. The close relationships and common research interest enabled many fruitful discussions. Gerd helped me to concentrate on the structure recognition and structure editing aspects.

I thank Jürgen Strauß for many discussions about a better structure of this thesis, Declan Mulcahy, Frank Buijs, and Peter Hennige for review the dissertation as well as Bernd Steinmüller, Hermann-Josef Kaufmann, Thomas Kern, Wolfgang Müller and other "Cadlabers" for their supports to this research. Further, I would like to thank Michael Tauber for introducing me into the domain of visual languages with many useful references.

Finally, I want to thank my wife Linfang for the support I needed to persevere, for her love and encouragement. I dedicate this work to her and our daughter Anja.

Paderborn, April 1993   Rui Zhao

Contents

List of Figures

List of Tables

Chapter 1

Introduction

The "computer-aided designer" of today can utilize CAD tools at every stage of the design process, from behavioral and functional specification to process simulation and optimization. Software engineers can make use of various CASE tools for program development. But the conceptual design is still usually done with pen and paper even when the designer has access to a powerful computer and is knowledgeable about working with it. Many resources of computers remain unused in this design stage. The extra transfer from paper into computer makes many primary ideas undocumented. It costs time and produces unnecessary errors.

One of the reasons that computers are not used at the first creative and conceptual design stage is that most of the current graphical interfaces are unfortunately "computer-centered" rather than "user-centered". The menu and command selection interface and the input devices mouse plus keyboard are not appropriate for conceptual sketching. The strengths of conventional user interfaces lie in the later stages of the design process, i.e. enter the finished design into computer, they provide only limited support for the early design stages. For drawing a rough sketch, or taking a short note, the interface is simply not as fast nor as convenient as pen and paper.

Conceptual design is usually supported by various diagram languages. Diagram languages are visual programming languages which use pictures formed from graphical elements as programs. Visual means using graphics instead of using text, because graphics can be easier comprehended and communicated by humans than text. One picture says more than thousand words. Graphics helps idea organization in conceptual design and communications in team project work. Examples of such diagram languages include many types of traditional diagrams used within computer science, such as Petri nets, statecharts, as well as graphical methodologies developed for software engineering, object-oriented analysis and design. Tools for drawing such diagrams are specific diagram editors, not general drawing editors.

Recently, notepad computers have become commercially available. The essential component that makes such computers attractive is the so-called "paper-like" interface which will emerge as a real alternative to the keyboard and mouse based one. An important advantage of pen-based computers is the mobility, it can be used everywhere. Such interfaces have several significant advantages which make gesture-based systems appealing to both novice and experienced users. A single gesture can specify a command with all required parameters simultaneously. A simple gesture can combine several commands in a natural manner. Diagram editing is an attractive application of gestural interfaces and pen-based computers which allow the user to communicate with computers in diagram languages by using handsketches so that the user can draw diagrams in the same way as with paper and pen.

While gesture-based diagram editors offer significant benefits, building such editors is difficult. Apart from the hardware improvement of flat displays with digitizers, the key problem is the gesture recognition which allows the user to sketch diagrams with relatively few restrictions. From the user's view, a gesture-based diagram editor should be modeless and intelligent to give the user a feeling that the editor understands the diagram language.

This thesis attacks the key software problems of the on-line gesture recognition and the integrated handsketch-based editor architecture. A novel incremental gesture recognition concept is presented and integrated in the object-oriented software architecture called Handi. Handi provides powerful programming abstractions for building handsketch-based diagram editors with less efforts.

1.1 Basic Concepts

As shown in figure 1.1, a diagram is represented in three layers: the internal representation in form of structured objects, the intermediate diagram in form of graphical descriptions based on graphical primitives like rectangles or lines, and the pixel-based picture on the screen inside a window for the human's eyes. Using diagrams directly with computers refers to the following human-computer communication problems: One is the automatic drawing of diagrams from the internal representation; another is the inverse problem, i.e. constructing the internal representation from a picture.

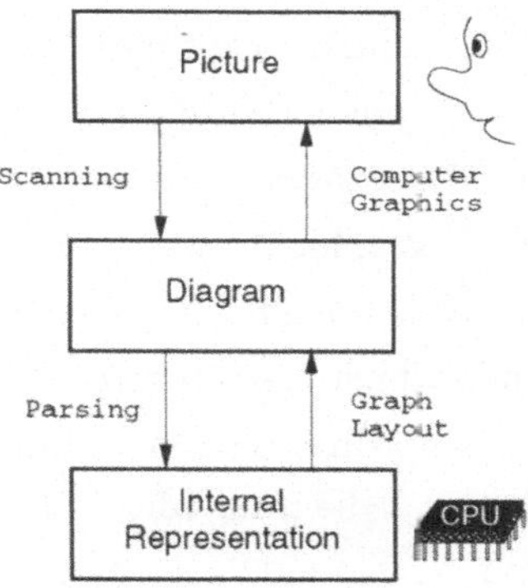

Figure 1.1: Basic problems in using diagrams with computers.

The problem to get a picture from a given internal representation refers to *graph layout* and *computer graphics*. The generation of a diagram description from an internal representation is the problem of graph layout. Computer graphics deals with the generation of display pictures from nonpictorial information. This work concentrates on the second problem, that is, from a picture to the internal representation of a diagram, which is clearly a pattern recognition problem. We classify this pattern recognition problem, again, into scanning and parsing. Mapping the external picture representation into primitive syntactical elements is the realm of *scanning*. Within conventional graphical user interfaces, command modes and direct manipulation techniques such as rubberbanding forces the scanning to a computer-centered style. Within gestural interfaces, scanning is an on-line recognition problem. To get an internal representation from basic syntactical elements is the realm of *visual parsing*. Similar to a textual program parser, a visual language parser depends on the underlying language syntax. Parsing a visual program is more difficult than parsing a textual program. This is because a visual program uses two-dimensional information to express its syntax and semantics. Recently, research in this domain concentrates either on handsketch recognition or on parsing visual languages. There has been a lack of integrated concepts and software architectures for building handsketch-based diagram editors.

Incremental Recognition

Within a handsketch-based diagram editor, the gesture recognition is a so-called
on-line recognition problem. On-line recognition means that the machine recognizes
pictures while the user is drawing. The input data of a gesture-based diagram
editor is a sequence of point coordinates captured by the input device. We call a
recognizer which transforms such point coordinates into graphical symbols, a *low-level recognizer* which refers to the scanner. The low-level recognizer determines the
class and the attributes of each graphical symbol drawn by the user. Further, a
gesture-based diagram editor needs a *high-level recognizer* to transform these basic
symbols into editing commands which are interpreted by a diagram editor in turn
to create the internal diagram structure. The most important issue is that the two
recognizers must work together in a diagram editor. Figure 1.2 shows the design of
our gesture recognizing system.

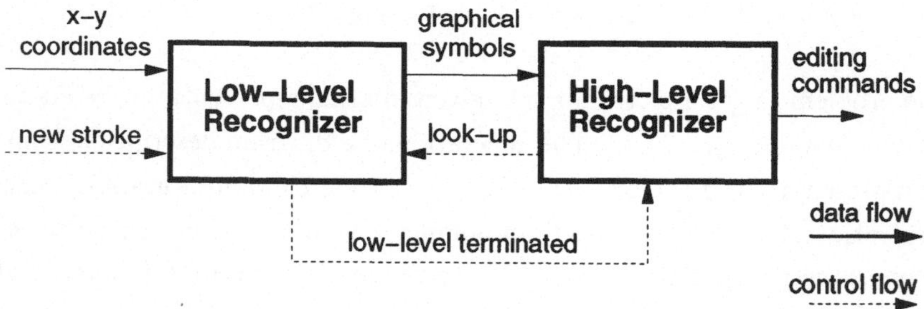

Figure 1.2: The incremental gesture recognition system

The essential idea of our incremental gesture recognition is to allow the high-level
recognizer *incrementally* transforming the graphical symbols recognized by the low-level recognizer into editing commands for creating and manipulating the underlying
diagram. This integrated recognition concept differs from existing approaches in the
following aspects:

1. Existing visual language parsers usually consider a complete picture as input,
   our high-level recognizer treats each graphical symbol the user has just drawn
   incrementally.

2. In contrast to other incremental visual language parsers which directly create the internal diagram representation, the output of our high-level recognizer are editing commands which are *compatible* with conventional diagram editors. This has the advantage that the gesture recognizer can be integrated into diagram editors which allow the user not only to draw new diagrams, but also modify existing diagrams with gestures.

Low-Level Recognition For solving the low-level recognition problem, a new method for on-line recognition of handsketched geometrical figures has been developed. In contrast to other on-line pattern recognition systems, strokes are not represented as feature vectors and matched to prototypes in a dictionary. Instead, a stroke is immediately classified top-down along a symbol hierarchy after it has been drawn. The recognized symbol will be displayed as a regular graphical object at once. A major advantage of this method is that the user can get an immediate response of recognition results. One significant feature of this novel method is that multiple-stroke handsketches can be recognized incrementally. This was a problem for most existing gesture-based systems. Object-oriented design has been used to build this low-level recognizer. A class-hierarchy of geometrical figures which makes use of inheritance is defined for encapsulating strokes and all recognizable geometrical objects. The polymorphism concept of object-oriented programming enables automatic and hierarchical control of the recognition process.

High-Level Recognition In contrast to the low-level recognition, the high-level recognition depends on the underlying diagram language. For this reason, we introduce the class of HiNet diagrams which mainly represent hierarchy and connectivity, and build a formal model of handsketch-based diagram editors. The key issue of the high-level recognition is how to specify the underlying diagram syntax. Our idea is to provide a mechanism to unify the specification of the language and its manipulations. We consider a visual language as an initial object and a collection of gesture editing operations. Any object that can be obtained by applying a sequence of allowed editing operations is then defined to be in the language. We specify the underlying diagram by defining a set of gestures, each one corresponds to an editing command. With our specification mechanism, each gesture defines a gesture shape, a set of gesture constraints, and the gesture semantics. The main task of the high-level recognizer is to check gesture constraints by examining the defined spatial relationship. One goal in designing the high-level recognizer is to permit an easy integration of a gesture recognizer into an object-oriented editor architecture. To achieve this goal, gesture semantics are defined as the generation of "normal" editing commands which can be interpreted by the underlying diagram editor in the same way as other commands.

Communication   Existing approaches merely identified a global data flow from a scanner to a parser, and therefore treated gesture recognition and diagram parsing as two separate and independent problems. The key issue of our concept is to consider them as two cooperative and tightly integrated components of the recognizing system of a gesture-based diagram editor. The cooperative communication between these recognition subcomponents supports our incremental gesture recognition, in which the user can sketch the desired diagram stroke by stroke, and the user immediately sees what happens after each stroke has been drawn. In order to achieve this incremental gestural dialog without any explicit command from the user such as "draw something, and click a button for parsing", our recognition system benefits by the inherent control signals which are illustrated by dotted arrow-lines in figure 1.2. Each pen-down event at the beginning of a new stroke produces automatically a "new stroke" signal which can be used to trigger the low-level recognition. The termination event of the low-level recognizer can be used to activate the high-level recognizer to parse graphical symbols into editing commands.

Handi Architecture

The basic design idea of Handi is to encapsulate common characteristics of hand-sketch-based diagram editors into classes by using object-oriented methodology. The concept of Handi is gained from experiences with several prototype editors. Handi integrates techniques of on-line handsketch recognizing, diagram parsing, and graphical structure editing into cooperative components of handsketch-based diagram editors. Handi consists of three subsystems: a sketching subsystem, a recognizing subsystem, and an editing subsystem. An editor for a particular diagram language relies on the sketching subsystem for handling free drawing input, on the recognizing subsystem for gesture recognition, and on the editing subsystem for its structure representing and editing capabilities. One of the key issue of Handi is to build Handi on top of a general editor framework by reusing the general graphical editing functional-

ity. Handi does not offer functionality which is supported by such editor framework
or any toolkits to avoid replicating existing functionality in Handi; instead we have
focused on providing new and previously unsupported capability, that is, the free
handsketching, the gesture recognition, and the creation and manipulation of diagrams.

1.2 Results and Contributions

This thesis solves the aforementioned software problems by an integrated concepts
for gesture recognition and an object-oriented software architecture for building
handsketch-based diagram editors. The primary contributions of this dissertation
which have partly been presented in [145, 146, 144, 148, 147] are:

- An integrated concept which combines the low-level recognition and the high-
  level recognition in an incremental recognition system.

- An object-oriented and hierarchical algorithm for on-line and incrementally
  recognizing handsketched graphical symbols. The main characteristics which
  distinguish our low-level recognizer from all existing recognition systems are
  an incremental control structure and a novel object-oriented architecture for
  efficient classification of geometrical figures.

- A formal model of HiNet diagram languages and handsketch-based diagram
  editors.

- A mechanism for gesture specification which integrates the diagram syntax
  definition and the editing operations.

- The Handi architecture with powerful programming abstractions for developing handsketch-based diagram editors.

- Two experimental diagram editors are built and evaluated, to demonstrate the
  viability of the basic concepts and the Handi architecture.

Thesis Organization

Chapter 2 discusses related work. Chapter 3 and 4 present the low-level recognizer and the high-level recognizer, respectively. Chapter 5 depicts the Handi architecture by using the Booch's graphical notation of class and object diagrams. Chapter 6 describes main aspects of a prototype implementation of Handi. Chapter 7 presents two experimental diagram editors. Finally, chapter 8 evaluates this work and chapter 9 summarizes the thesis and discuss directions of future work.

Chapter 2

Related Work

In this chapter, we briefly state the relationship of this thesis to similar work in the related field classified into four categories: gestural interfaces, visual language systems, pattern recognition systems, and graphical structure editors. The general relationships are as follows: A handsketch-based diagram editor is a specific application of gestural interfaces. Diagrams are visual languages, and diagram processing relates closely to concepts of visual programming systems. The key problem of handsketch-based diagram editing is the on-line gesture recognition which is a pattern recognition problem. A diagram editor is a graphical structure editor with specific input technique.

2.1 Gestural Interfaces

Some general work attempts to define gestures as a technique for interacting with computers. Morrel-Samuels [81] examines the distinction between gestural and lexical commands, and then further discusses problems and advantages of gestural commands. Wolf and Rhyne [138] present a taxonomy of direct manipulation which considers gestural interface as a member of direct manipulation interfaces. Baecker and Buxton [8] discuss human factors concerning gestural interfaces, as well as hardware and software issues. Buxton studies the lexical and pragmatic considerations of input structure [14], specially for performing selection and position tasks [13], and discusses the use of muscular tension and motion to phrase human-computer dialogues [15, 16].

2.1.1   Notepad Computers

The idea of using gestures and handwriting to interface with computers has attracted people for many years. The graphics tablet and stylus have been in use since at least 1964 [25]. Interactive graphics displays have been in use since at least 1963 [116]. Despite the existence of these tools, communication with stylus and display bears little resemblance to the way we communicate with pencil and paper, chalk and blackboard. We all write letters, understand basic proofreading symbols, and software engineers discuss their designs with various block diagrams. But with very few exceptions [130, 57], today's user interfaces make little or no use of these skills.

Recent advances in devices and VLSI technology have been used to realize a notepad computer in the size and the weight of a book. A few of commercial pen-based computers which utilize character and gesture recognition techniques become to be available, for example, NCR's NotePad, Momenta's Pentop, IBM's Thinkpad, GRiD's Pad Pad [127]. Many pen-oriented operating systems and window systems are getting available, for example, GRiD's PenRight!, GO's PenPoint[17, 90], CIC's PenDOS, and Windows for pen computing [21].   Altabet [3] discussed an integration of pen-based computer and multimedia technology to provide a truly natural human interface.

2.1.2   Applications

In the early seventies, Alan Kay described the idea with a so-called Dynabook [58] which may be considered as the first approach for building pen-based computers. Unfortunately, only a little work was dedicated to designing and developing pen-based computers and gesture-based systems due to the relative low advanced digitizing technology and difficulty of handwriting and handdrawing recognition. In [113], early problems and issues were discussed which limited the acceptance of user interfaces using gesture input and handwriting character recognition. Nevertheless, some early work has been done by using handwriting interfaces or gestural interfaces in different application areas.

Hosaka and Kimura [50] used handwriting input in an interactive geometrical processing system for designing and manufacturing of three dimensional objects. A graphic tablet digitizes the user's handwriting which can be recognized. Many function keys are used in this early work to allow the input process with a tablet. For example, the user must press a function key to begin a drawing process.

Odawara et al [92] presented a design environment for silicon compilation by using a LCD digitizer in the same style as today's pen-based computers. A diagrammatic hardware description language (ADL) forms the input of this silicon compiler. The designer can draw ADL diagrams like drawing on paper, and therefore is able to concentrate his attention upon the design for a long time. The system can recognize handwritten characters and ADL symbols.

Gestures are also used in architecture design. Makkuni [75] developed a system which allows a user to design Chinese temples with gestures. Makkuni described a gesture language which supports the design process such as gesturally exploring a pavilion roof.

Jackson and Roske-Hofstrand [53] used circles as select gestures for mouse-based selection without button presses. Circling motions are detected automatically; their experiments show that many users prefer circling over button clicking for selecting objects.

At IBM, many research has been done in the *paper-like interfaces* project [102]. The goal of this project was to develop a body of knowledge about the applicability of gestural human computer interfaces, and to explore software technology for the development of gestural interfaces. Within this project, Rhyne et al discussed the dialogue management for gestural interfaces [101] and described a prototype electronic notepad [100]. Wolf et al presented several prototype applications of the paper-like interfaces such as information processing with spreadsheets [102], educational applications [20], medication charting [4], freehand sketching, gestural creation of music score and interpretation of handdrawn mathematical formulae [139], and support of group work [137]. They presented several analysis on how well such interfaces work [135, 136].

At MCC, the *Interactive Worksurface Projects* have been completed for building CAD systems using handwriting recognition [49]. Successful research has been done in interactive tablet hardware [6], handwriting recognition with neural networks [77, 76, 96], and visual languages [134]. Our goal is similar to that of MCC, however, we concentrate on the handsketch-based diagram editing within a graphical structure editor. In contrast, MCC mainly investigated hardware design and fundamental research in handwriting recognition with neural networks.

Kurtenback and Buxton [68, 67] designed a prototype graphical editor (GEDIT) that permits a user to create and manipulate three simple types of objects using shorthand and proofreader's type gestures.   Using handdrawn symbols, the user adds, deletes, moves and copies these objects. The most essential difference between GEDIT and Handi-based editors is that Handi-based editors are graphical structure editors for diagram languages, and GEDIT is a very primitive general drawing editor just for pictures consisting of squares, circles, and triangles.

Furthermore, gestures are also used in combination with natural language processing for multimodal reference. Schmauks and Reithinger [110, 2] discussed the application of *pointing gestures* in natural language dialog systems.   However, Schmauks used gestures mainly for pointing, which differs from our sketching-oriented gestures. Gestures were classified into punctual pointing gestures and non-punctual pointing gestures.

GRANDMA

The gesture-based system GRANDMA [106] developed by Dean Rubine comes closest to our research. Rubine describes two methods of integrating gestures and direct manipulations. First, GRANDMA allows views that respond to gestures and views that respond to click and drags to coexist in the same interface. Second, GRANDMA supports a new two-phase interaction technique, in which a gesture collection phase is immediately followed by a manipulation phase, which is called eager recognition [108].

Similar to GRANDMA and different to several other approaches, Handi-based editors support the *coexistence* of the two interface techniques, that is, within Handi-based editors, the user can still click, drag, and rubberband all graphical objects. Differing from the eager recognition approach, Handi supports an incremental sketching style which is more appropriate for editing diagrams. However, the immediate feedback of our low-level recognition has the same goal as the eager recognition, that is, to avoid that an entire gesture must be entered before the system responds. In GRANDMA, only single-stroke gestures can be used, in a Handi-based editor, there is no restrictions in the stroke number of gestures. Further characteristics which differ Handi from GRANDMA are: 1) GRANDMA is built from scratch by directly using the X window system, Handi is built on the top of a general editor framework. 2) Handi comprehensively supports the development of handsketch-based diagram editors. In contrast, GRANDMA concentrates on the input model with sophisticated event handlers. 3) Handi supports multiple-*stroke* gestures, GRANDMA supports multiple-*path* gestures of multiple-finger input [105]. 4) GRANDMA supports *irregular* gestures which are represented as a vector of real-valued features, and its recognizer must be trained by the user with many examples. Handi supports *geometric* gestures, and provides an extensible set of frequently used graphical symbols which can be used without training. 5) In GRANDMA, gesture semantics are specified in form of gesture interpreters which manipulate the internal objects directly. In Handi, gesture semantics are specified in form of editing commands.

2.2   Pattern Recognition

Apart from the hardware problems, the main barrier for wide usage of gestural interfaces and pen-based computers is the on-line recognition of handsketched gestures and handwritten characters. The state of the art in on-line gesture and handwriting recognition may not be well-known outside its particular field. It has been a topic covered more often in pattern recognition community than in user interface design and computer graphics. Tappert et al [122] provides a comprehensive survey of the different approaches taken to on-line recognition.

2.2.1   Character Recognition Systems

The development of electronic tablets in the 1960s led several researchers to attempt the on-line recognition of handwritten characters. Some of these early attempts were rather successful [88], but the interest gradually diminished. Recently, there has been a resurgence of on-line pattern recognition due to the appearance of pen-based computers, high performance graphical workstations, and national language considerations (Chinese and Japanese character input). There are many systems designed to recognize different types of characters, for example, digits [24, 27], English [96], Arabic [28], Chinese [140], Japanese [62] letters, and mathematical symbols [10, 26].

Decision Tree Kerrick and Bovik [59] designed a microprocessor-based character recognizer which is closely related to the hierarchical control structure of our low-level recognizer. A binary decision tree with simple features is used to rapidly reduce the set of candidate characters to a very small set. However, our hierarchical classification of graphical symbols is not the same as a decision tree for the following reasons: All nodes in a hierarchy of geometry represent reasonable recognition results. On the contrary, in a decision tree, only leafs represent recognition results. All nodes in a decision tree represent ambiguous states. A decision tree is a binary tree, a symbol hierarchy is not restricted to be a binary tree. Decision trees are used only for classification of characters, our hierarchical recognition combines classification and feature analysis with automatic control in terms of *object recognizes itself*. Further, our low-level recognizer calculates features only when they are required, while within a decision tree classification all features are calculated before any classification is initiated.

2.2.2 Handsketched Figure Recognition Systems

There are only a few approaches investigated to on-line recognition of handsketched figures. Murase [82] describes a system for recognizing handsketched flowcharts. However, his method is designed for recognizing complete flowcharts, that is, the user has to draw a complete flowchart before the recognition can begin. All subfigures that can be symbols are extracted from the input sketch. Elastic matching distances are calculated between these candidate symbols and prototype symbols. Finally, the system simultaneously recognizes and segments the entire figure by choosing the candidate sequence that minimizes the total sum of distances along the sequence. Although this system considers a similar picture class, the design philosophy is total different. This recognition system is not designed for a gestural interface, rather the user has to draw a complete diagram and then start the recognition system.

Similar to Murase's approach, Kojima and Toida [63] developed a system for on-line handdrawn figure recognition. An adjacent strokes structure analysis method (ASSAM) is described. The figures are classified into fundamental figures and symbols. A fundamental figure means a line segment or a closed figure composed of only one loop. A symbol means a figure composed of several fundamental figures. The recognition algorithm is composed of two steps: fundamental figure recognition and symbol recognition. The fundamental figure recognition is done by analysis

of the number of apexes and the categories of line segment between them. While
the recognition algorithm of fundamental figures appears to be quite ad hoc, the
approach of combining of adjacent strokes makes the algorithm independent of the
stroke-order and stroke-number.   The combination of adjacent strokes is similar
to our approach of the incremental updater. The essential difference is that our
low-level recognizer combines connected strokes into a new stroke which is displayed
immediately on the screen. However, the combination used in ASSAM is merely a
technique of the matching algorithm.

2.2.3   Gesture Recognition Systems

Gestures have properties that are different from those of handwritten characters.
For example, gestures do not have regular heights and orientations. Therefore, new
recognition methods for gestures are necessary.

Dean Rubine [107] presented a trainable statistical gesture recognizer for single-
stroke gestures. The recognition is done in two steps. First, a vector of features
is extracted from the input gesture. The feature vector is then classified as one of
the possible gestures via a linear machine. The intelligence in Rubine's algorithm is
in its many characteristics, 13 in all, which characterize a stroke. In his algorithm,
the usage of weights for specified characteristics is similar to neural net weight
adjustment. The weights are determined by training from example gestures.

Kim [60] presented a gesture recognizer based on feature analysis which has been
improved and redesigned by Lipscomb [73] who combines techniques of angle filtering
and multi-scale recognition. An angle filter is used to reduce noise and quickly distill
the many input points of a stroke. The angle filter produces output points where
stroke curvature is high, simply said at the corners. Later, a recognizer uses a feature
finder to decide which candidate features are significant. These features match a
stored prototype stroke, triggering recognition. Recognition succeeds only when the
known and the unknown stroke have the same number of points. This is achieved by
training. In contrast to Rubine's algorithm, the multi-scale recognizer concentrates
its intelligence in its multi-scale data structure, not in its stroke characteristics or
weighting.

A common feature of these gesture recognizers is the usage of training which is
good for gestures used only for a single application, but not appropriate for a class of

diagram languages. In section 3.2.3, we discuss some further differences between an irregular gesture recognizer and a symbol recognizer after our problem is analyzed in detail.

2.3   Visual Languages

The term visual language is used to describe several types of languages: languages manipulating visual information, languages for supporting visual interactions, and languages for programming with visual expressions [112]. Myers [84] emphasizes the difference between visual programming and program visualization systems. Visual programming refers to any system that allows the user to specify a program in a two-(or more)-dimensional fashion. Program visualization is an entirely different concept from visual programming. In visual programming, the graphics are used to create the program itself, but in program visualization, the program is specified in a conventional, textual manner, and the graphics is used to illustrate some aspect of the program or its run-time execution.

In this work, we consider only visual programming languages which may be further classified, according to the type and extent of the used visual expressions [112], into icon-based languages, form-based languages, and diagram languages which are the target languages of this thesis.

2.3.1   Specification and Parsing

Visual programming environments have been a research topic for many years, there are different approaches for the definition and parsing of visual languages. Spatial parsing is the process of recovering the underlying syntactic structure of a visual program from its spatial arrangement. Most of the existing approaches of spatial parsing are grammar-based and batch-oriented. Examples of visual language grammars are picture layout grammar [39], positional grammar [22], relation grammar [23, 31], graphical functional grammar [64], unification-based grammar [134], and constrained set grammar [47]. A grammar-based visual language parser is designed to be generated from a grammar definition. Its user interface is similar to that of a conventional program compiler. The input is a picture in a certain format such as a picture description [40], PostScript [54], or bitmap [134]. The output of the parser consists of a statement about the syntactical correctness of this picture and an attributed parse structure. However, the visual programming environment which uses such a parser has the same style as the currently used textual programming environment: the programmer has to draw a complete diagram and input it to the parser. One serious problem is the visualization of error messages of spatial parsing, because errors of a visual program cannot be reported by using line numbers like conventional compilers do. The batch-oriented approach of spatial parsing is obviously not appropriate for interactive editing.

Similar to the spatial parsers, our high-level recognizer recognizes handsketches from the spatial arrangement as well. However, the high-level recognizer is designed as an integrated component of handsketch-based diagram editors, so that the parse structure is at the same time the internal object graph of the structure editor. The most important issue of our high-level recognition differing from existing spatial parser approaches is the need to process input incrementally while imposing no constraints on the input order of the graphical symbols produced by the low-level recognizer.

A Constrained Set Grammar [47] is a collection of productions which consist of a set of non-terminals on the left-hand side, a set of symbols on the right-hand side, and a collection of constraints between the symbols on both sides. Constraints are the key feature of the constrained set grammar which enable information about spatial layout and relationships to be encoded in the grammar. The spatial relationships used in the constrained set grammar are similar to those used in the gesture constraints of our high-level recognizer. However, the constraints are checked with totally different techniques. A constrained set grammar is transformed into a set of clauses written in the constraint logic programming language CLP [46], therefore, logic programming tools for specifying constraints as well as general-purpose theorem provers can be used. However, the computational cost incurred by such methods is so high that such parsers are very slow.

Unification-based Grammar designed by Wittenburg et al [134] supports parsing handsketched inputs from a graphical tablet. Their goal corresponds closely to our goal, that is, recognizing handsketched diagram pictures. A unification-based grammar and a parsing algorithm are presented for defining and processing visual languages. The lexical lookup process is represented as a set of productions that maps exclusively from terminals to nonterminals. Two testbed applications for the parser are implemented, a math sketchpad and a flowchart sketchpad, which recognize mathematical expressions and structured flowcharts, respectively. These systems were targeted for the MCC's Interactive Worksurface discussed in section 2.1. A feature of these systems similar to Handi-based editors is the possibility to accept elements in the order they are drawn, rather than in some spatially defined ordering. However, Handi-based applications are structure editors both for incrementally creating diagrams and manipulating existing diagrams. Another significant feature of the Wittenburg's system is that it collects strokes until the user exceeds a time-out threshold between strokes. Further, the set of strokes are presumed to represent a single symbol of the input vocabulary. In contrast, our incremental recognizer immediately recognizes each input stroke, which does not force the user to make explicit pauses between strokes of different symbols.

2.3.2 Visual Programming Systems

SIL-ICON is a visual compiler [18] developed at the University of Pittsburgh which supports the specification, interpretation, prototyping, and generation of icon-oriented systems. An icon interpreter uses a formal specification of an icon system to understand and evaluate a visual sentence. The design of the system is based on the concept of a *generalized icon*. A generalized icon has a dual representation (X_m, X_i) where X_m is the logical part (meaning) and X_i is the physical part (image). An essential characteristic of the generalized icon concept is that the logical part and the physical part are mutually dependent, which is similar to Handi's concept of dual representation of diagrams in components and views. In Handi-based systems, gestures always relate to the diagram's view, in the same way as icon operators with the physical part of the icons.

The physical part of an icon is specified by a picture grammar. A picture grammar is a context free grammar where the terminal symbols include both primitive picture elements and spatial image operators. The operators describe compositions of the physical parts of icons. SIL provides three operators: horizontal concatenation (noted with the character '+'), vertical concatenation ('∧'), and spatial overlay ('&'). Using these operators, a string can describe a complex physical icon. For example, the iconic sentence in figure 2.1 can be represented by the string

$$(box + box) \& cross$$

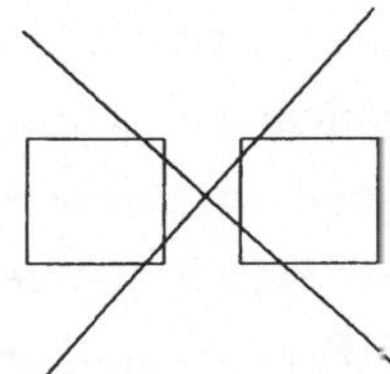

Figure 2.1: Iconic sentence

GREEN is a GRaphical Editing ENvironment designed by Golin [40] which allows the programmer to create and manipulate a visual program. The visual structures can be recovered by a visual parser. The language syntax is defined by the picture layout grammar [39]. An editor provides the user interface to create a set of primitive picture elements in a picture file which can be processed by the parser. The parser tries to find a valid parser structure among all possibles by using multiple set structure. The set of graphical primitives is restricted to boxes, octagons, circles, text, lines and arrows with fixed attributes. The performance of this system is too slow to be practical for use as a program development system [40]. The editor is not designed to allow the user to make handsketches, instead a grid alignment mechanism is provided to guarantee that attached objects have appropriate coordinates.

2.4 Graphical Structure Editors

Graphical structure editors is a subdomain of *visual programming environment* which is a software tool or collection of tools to support programming in a visual language. The role of the visual programming environment is analogous to the role filled by a traditional, text-based programming environment such as Cornell Program Synthesizer [124]. Visual Programming environments must support two basic tasks: the creation and manipulation of visual programs; and the processing of visual programs by analying and executing them. *Diagram editing systems* refer to the first one which again consists of several different aspects. Systems for *graph drawing* such as Edge

[87] or *diagram visualization* such as Compound Digraph [115] consider aspects of aesthetics layout of graph or diagrams which are not considered in this work.

This work relates to graphical structure editors because handsketch-based diagram editors are a kind of graphical structure editors. A number of graphical structure editors have been designed with several different aspects which can be classified into two basic principles. One is the generator approach such as GEGS [117], PAGGED [38], and LOGGIE [7]. Another is the toolkit approach such as Unidraw [129], ET++ [131], and Diagram Server [9].

The essential difference between the generator approach and the toolkit approach is the manner of how to reduce the development efforts of structure editors. Within the generator approach, the editor designer specifies an editor with a grammar, and an editor can be generated. Within the toolkit approach, basic building blocks with desired functionality are provided as reusable objects and classes so that an editor can easily be developed. This work follows the toolkit approach because there are so many common features among HiNet diagram editors, which can be encapsulated into reusable classes.

The common characteristics of all graphical structure editors are that graphs are used as internal representations. These graphs are characterized by user-defined node types and edge types. Some systems allow constraints such as "every node of type x must be connected to at least one node of type y." Considering the user interfaces of conventional graphical structure editors, all systems include sophisticated commands with many complex modes for entering and deleting picture elements. These commands can usually be selected from a language-dependent palette of command buttons or from menus. Language-independent commands like file operations, zooming, and scrolling are standard-commands.

Similar to graphical structure editors, the internal representation of the underlying diagram structure within our high-level recognition is an object graph as well. However, we use the object-oriented methodology to represent different node types and edge types in appropriate object classes. Basic concepts for representation and recognition of hierarchy and connectivity are developed. In contrast to conventional structure editors, handsketch-based editors use gestures to invoke commands for entering and deleting graphical elements. The input is in free form, modeless in any order and place. The user can express ideas directly by handdrawing. Therefore, command palettes are usually not necessary. For example, in a handsketch-based

Petri net editor, creating a place object can be done by drawing a circle; and creating a transition object can be done by drawing a rectangle.

GEGS The Graphical Editor Generator System designed by Szwillus [117, 118, 119] is an editor generator system with a powerful specification method based on attributed grammars which can specify four categories of information: 1) a class of directed graphs with node and edge types as the set of valid internal editing objects, 2) graphical presentations associated to graphical node types, 3) rules and dependencies between nodes and edges of the internal structure graph, 4) different editing modes to allow switching constraints checking on and off. The basic idea of GEGS is to generalize the concepts of textual structure editing to graphical languages. A significant feature of GEGS is that the combination of visual language specification and the user interface specification in one language. This is not at all dissimilar to our mechanism of the gesture specification which combines gesture shapes, gestures constraints, and gesture semantics. However, our gesture specification is not considered as components of a generator input language, and GEGS does not consider gestural interfaces at all.

Unidraw designed by Vlissides [129, 128] comes closest to the basic principles used in our Handi architecture. A part of the current Handi implementation is built on top of Unidraw. Unidraw simplifies the construction of graphical editors by providing programming abstractions that are common across domains. Unidraw defines four basic abstractions: components encapsulate the appearance and behavior of objects, tools support direct manipulation of components, commands define operations on components, and external representations define the mapping between components and the file format generated by the editor. Unidraw emphasize the generality of such an editor framework, and supports a broad range of domains such as technical and artistic drawing, music composition, circuit design, and many others. Due to the endeavor of the generality, a diagram editor designer still must implement many common features among diagram editors. Further, Unidraw does not consider the input technique of free hand drawing.

Chapter 3

Low-Level Recognition

The input medium of a handsketch-based diagram editor is a sequence of point coordinates captured by the used input device. A low-level recognizer transforms such point coordinates into graphical symbols. We begin this chapter with the problem analysis, consider several related pattern recognition problems, and state why these existing methods are not appropriate for solving our problem. Then we present our fundamental concepts and design decisions which lead to an object-oriented system design. We describe the system components and algorithms in detail and conclude the chapter with a summary of our method.

3.1 Problem Analysis

3.1.1 Requirements

Handsketch-based diagram editors are specific gesture-based systems which place several specific requirements on the low-level recognition. The requirements we identified are as follows:

1. Recognition must be fast. Response time is widely acknowledged to be one of the chief determinants of user interface [111, 8]. Response time in direct manipulation systems is especially important as noticeable delays destroy the feeling of directness. The recognition results must be seen by the user immediately after a stroke is drawn.

2. Recognition must be activated automatically. This means that the recognition system should start the recognition process after the user has drawn a stroke, without any explicit demands of user's commands. A dialog style such as "draw a gesture and then click a command button to recognize this gesture" reduces the directness of gestural interface and is therefore not acceptable.

3. Multiple-stroke gestures must be recognized. The user usually sketches in a multiple-stroke style as that used with paper and pen, which should be supported to make the handsketch-based user interface more natural. Further, some symbols can't be drawn in a single stroke, that means, multiple-stroke recognition is necessary on using such symbols as gesture.

4. Recognition should be robust and tolerant. The recognition rate in an on-line recognition system is strongly dependent on the care of the user's drawing. However, it is also important in conceptual design to allow the designer making hasty sketches which can still be recognized by using the underlying diagram syntax.

5. The recognizer should be versatile and extensible. The recognizer is designed for a class of diagram languages not for a single diagram language with a fixed set of symbols. Therefore the recognizer should provide on the one hand a large set of basic geometrical symbols, and on the other hand it should be easily extensible for new symbols which are not in this basic set.

3.1.2 Input of Handsketches

The input device used for a gestural interface can be a digitizer pen or a mouse which generates point coordinates of each handsketch. *Stroke* and *inking* are the two most important terms frequently used within gestural interfaces. A stroke is the drawing from pen down to pen up, which is originally represented by a sequence of digitized point coordinates. Inking is a technique widely used to immediately show the digitized data to simulate paper and pen. This can be done in two different styles:

- draw each digitized point,

- draw a polygon whose vertices are the digitized points.

Figure 3.1 is a screen dump of the inking of a handdrawn rectangle by using the second inking technique. The white breaking points are the digitized points. Figure 3.2 shows the captured x-y coordinates of this handsketched rectangle which is drawn in a single stroke. Twenty points are digitized by this handsketched rectangle. Parameters such as a minimum distance between two adjacent points can be defined as an input filter to avoid receiving irrelevant points. For example, in using digitizers which send point coordinates regardless of the pen movement, the input filter is necessary to omit coordinates received which belong to the same point.

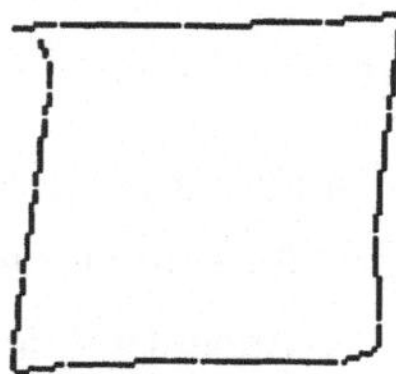

Figure 3.1: A handsketched rectangle

Nr.	x	y	
0	184	389	-- pen down
1	194	390	
2	222	390	
3	254	391	
4	272	393	
5	269	371	
6	266	343	
7	267	323	
8	267	317	
9	258	314	
10	226	314	
11	196	313	
12	184	311	
13	184	317	
14	186	333	
15	189	353	
16	191	369	
17	192	375	
18	192	382	
19	190	386	-- pen up

Figure 3.2: The point coordinates of a handdrawn rectangle stroke

3.1.3 Specific Properties and Problems

Compared to off-line picture recognition, on-line pattern recognition has the advantage that the segmentation is not a problem because segmentation can be done automatically while digitizing. Handsketches are segmented usually in strokes by using the pen-down and pen-up information. Strokes are stored in arrays of x-y coordinates in the same order as they are digitized. However, the representation form of arrays of x-y coordinates brings several problems which are not present in applications of off-line picture recognition. These problems appear both in single-stroke drawings and multiple-stroke drawings.

3.1.3.1 Single-Stroke

One single-stroke recognition problem is that there are many different possibilities for drawing a certain geometrical figure. Figure 3.3 illustrates this problem by giving examples of different single-stroke rectangles.   With the restriction that the start

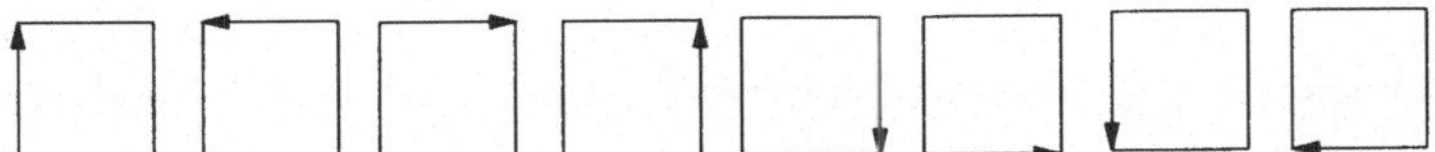

Figure 3.3: The eight variants to a single stroke rectangle

point must be at one of the four vertices of the rectangle, there are eight variants to draw the same rectangle by choosing different drawing directions and start points.

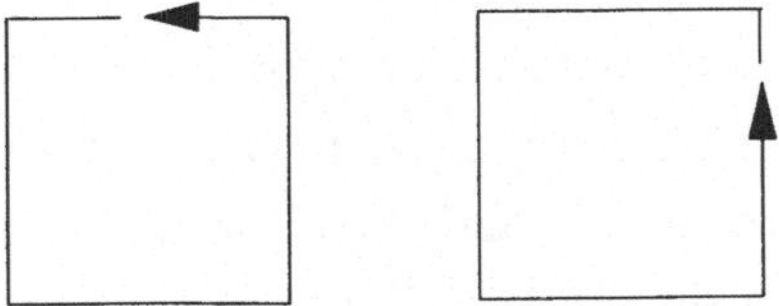

Figure 3.4: Start at the edge to draw a rectangle.

If allowed to start at any position as shown in figure 3.4, the number of drawing variations is unlimited. In the representation form of on-line captured arrays of the x-y coordinates, these are all different "patterns" which makes the recognition task quite complex.

3.1.3.2 Multiple-Stroke

Most geometrical figures are usually drawn by using several strokes instead of a single stroke. For example, even if giving the restriction that an edge of a rectangle should be drawn in a single stroke, the same rectangle can be drawn in five different combination forms as shown in figure 3.5.

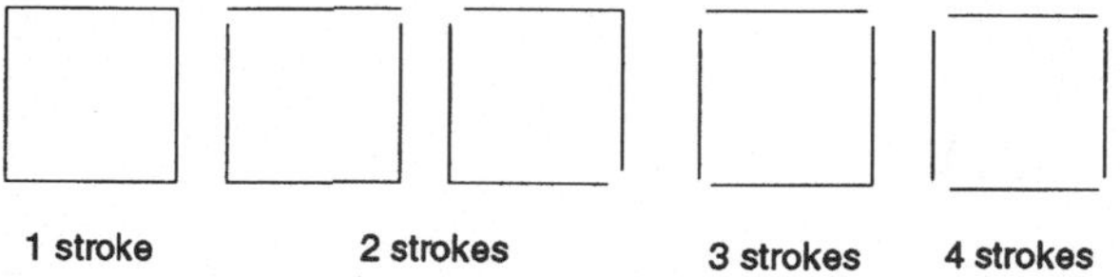

Figure 3.5: There are five stroke combinations of a multiple-stroke rectangle.

Unlike characters, geometrical figures do not have well-defined stroke orders. Stroke-orders are often used to reduce the computational costs in on-line character recognition, particularly in recognizing Chinese characters. In contrast, an on-line geometry recognizer must be able to handle all possible stroke-orders for a given figure. The temporal information of stroke-order, which brings benefits for on-line character recognition, is rather a problem in recognizing graphical symbols.

In figure 3.3 and 3.4 several variants were illustrated for drawing a single-stroke rectangle. Different start positions and different drawing directions generate different patterns. There are much more variants for drawing a multiple-stroke rectangle than for drawing a single-stroke rectangle. In case of drawing a four-strokes rectangle by considering the stroke-order and the drawing directions of each stroke, there are $16 \times 4! = 384$ different drawing possibilities. Together with five different stroke combinations, there are several hundreds of "patterns" for rectangles. Figure 3.6 gives four reasonable examples among them.

The examples above deal exclusively with rectangles. It is clear that the same problem exists in drawing other geometrical figures as well. Figure 3.7 illustrates this by giving a few other examples.

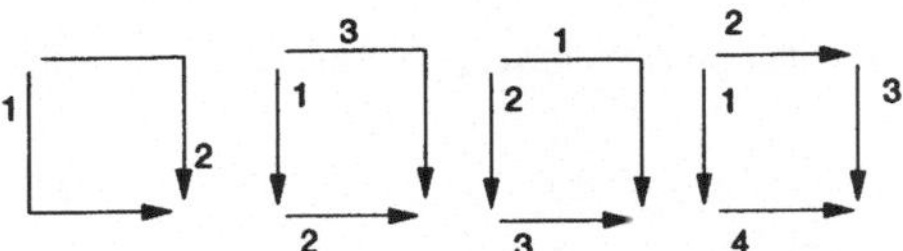

Figure 3.6: Considering stroke-order and drawing directions

This multiple-stroke problem is one of the most difficult problems of on-line pattern recognition. Most of the existing gesture recognition systems provide only single-stroke recognition. Other systems consider this problem in the gestural dialog manager [73]. In systems which do not provide immediate response such as in [134], multiple-stroke problems are treated as follows: At the input level, strokes are collected until the user exceeds a time-out threshold between strokes. The resulting set of strokes, which must represent a single symbol, is sent to an image recognizer, e.g. a neural net, for classification. However, such a system cannot recognize graphical attributes of a symbol.

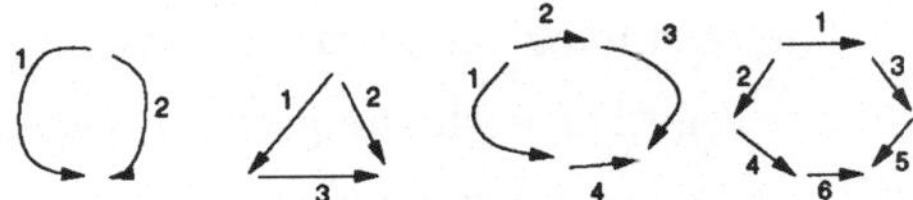

Figure 3.7: Variants of drawing various geometrical figures

3.2 Related Problems

3.2.1 Overview

Pattern recognition has a long history, many methods have been developed for various applications. Because of its practical importance, pattern recognition has been a very active field. Various approaches to different applications are developed. The most important system condition of a pattern recognizer is the form of the input pattern and how these patterns are captured. Pavlidis [95] distinguishes four classes of pictures: full gray scale and color pictures, bilevel pictures, continuous curves and lines, and points or polygons. The forms of pictorial data introduce pattern recognition applications into two research fields: on-line and off-line recognition.

Off-line Recognition   Traditionally, pattern recognition considers mainly the off-line recognition problem, that is, the automatic construction of symbolic descriptions for pictures which are color pictures or bilevel pictures. The input is usually scanned with all the background information. Segmentation, contour tracing, thinning or scene analysis are the main problems within off-line pattern recognition. One of the most difficult problem of off-line drawing recognition is the segmentation problem, that is, the scanned data must be converted to line drawings. This requires costly and imperfect preprocessing to extract contours and to thin or skeletonize them. Typical examples for off-line pattern recognitions are: classification of OCR characters [12], recognition of handdrawn schematic diagrams [93] and mechanical drawings [52]. The term off-line signifies that the recognition is performed after the picture is created.

On-line Recognition   The technological developments of the last years have made computer graphics popular. Together with the growing performance of computers, on-line pattern recognition has been become interesting. In contrast to off-line recognition, on-line recognition means that the machine recognizes handsketches while the user is drawing. The input data are vectors of x-y coordinates obtained in real-time from a digitizer pen or a mouse. Therefore, on-line recognition deals with pictures of the points classes. "Electronic ink" displays the trace of the drawing on the screen, and recognition algorithms instantly convert the coordinate data into appropriate symbolic descriptions. Recently, research in on-line pattern recognition has focused on handwritten characters [59, 28, 122], excepting some research into gesture recognition [75, 73, 107]. These recognition problems are all on-line recognition problems which consider a common picture class of points or polygons. The input patterns in on-line recognition are structured in strokes which vary in both their static and dynamic properties. Static variation can occur, for example, in size or shape. Dynamic variation can occur in stroke number and order.

Common characteristics can mislead to use methods for character recognition directly to recognize handsketches. However, through detailed analysis of these problems, it is confirmed that they are quite different problems. As mentioned in

[8], one of the problems with on-line pattern recognition is that we tend to lump all of the different approaches together. In fact, there is probably as much stylistic difference between a system that recognizes blockprinted characters and one that recognizes proofreading gestures as there is between a menu system and one that uses direct manipulation. Although there are many similarities between character recognition, irregular gesture recognition, and geometry recognition, the methods for character recognition and irregular gesture recognition cannot be used directly to recognize handsketched geometrical figures. This is because there are several significant distinctions between these recognition problems, each of which relates to different requirements. We want point out what are the common features and where are the main differences to prepare our design decisions.

3.2.2 Character Recognition

Character recognition has a long history, and a number of character recognition systems have been developed. The state of the art of on-line character recognition is surveyed in [122]. The advent of electronic tablets in the late 1950's precipitated considerable activity in on-line handwriting recognition. This intense activity lasted through the 1960's, ebbed in the 1970's, was renewed in the 1980's, and has become popular now. The renewed interest in on-line character recognition stems from the advent of notepad computers.

Pattern Classification and Pattern Analysis   Niemann [89] states that pattern recognition comprises classification of simple patterns and analysis of complex patterns. A pattern is considered to be simple if a class name is sufficient, and classification means that each pattern is considered as one entity and put into one class out of a limited number of classes. No quantitative characterization is attempted in classification. A pattern is considered to be complex if a class name is not sufficient. Analysis of a pattern means that an individual description of each pattern is given. Therefore, quantitative characterization is usually necessary in pattern analysis. Handsketched strokes are complex patterns because a class name is not sufficient and the classification of a stroke as a whole is not feasible. Strokes differ from each other not only in class names but also in quantitative characterizations such as positions and dimensions as discussed in section 3.1. It is apparent that there is an overlap between pattern classification and pattern analysis.

The most important reason why character recognition methods cannot be used directly in geometry recognition is:

Character recognition is only classification. Geometry recognition is not only classification but also analysis of graphical attributes.

In contrast to character recognition, it is not enough *just* to recognize the class which a handsketched geometrical figure belongs to, the graphical attributes must be recognized as well. Therefore, a geometry recognizer should recognize the shape-type of a handsketch, and at the same time it must identify all graphical attributes such as size and position of feature points. This is because all of this information are used to define a visual language syntax. The relationship between the geometrical objects, the size and shape of them have both syntactical and semantic meanings. Considering the handsketched rectangle shown in figure 3.1, a geometry recognizer should output at least the following information:

1. it is a rectangle

2. this rectangle is upright.

3. the coordinates of the low-left and the upper-right vertices of this rectangle are (184, 311) and (272, 393).

Many efforts in character recognition are faced with the problem of shape discrimination between characters that look alike such as U-V, C-L, a-d, n-h, O-0, 1-l, Z-2 [122]. Another difficult problem in character recognition is cursive writing recognition [121, 33]. Recently, new research results show that neural nets have excellent performance on solving such problems [96]. But neural nets cannot output information such as vertices-coordinates of geometrical figures. They are good in classification but poor on analysis.

While character recognition treats a large set of characters, geometry recognition considers a relative small number of essentially different geometrical shapes but with an infinite number of hierarchical parameter variations. In contrast to character recognition, the main characteristic of geometry recognition is to identify feature points of handsketched figures.

3.2.3 Irregular Gesture Recognition

Gestures for editing diagrams are regular geometrical figures like rectangles, circles
or lines. Such gestures are called *regular* gestures. We use the term *irregular gesture*
to indicate hand markings which do not have regular shapes compared to geometrical figures. Such gestures are usually designed only for certain applications. Figure
3.8 illustrates four examples of such gestures: (a) delete-gesture used in [73], (b)
ellipse-gesture in [107], (c) merge-paragraph-gesture in [56], and (d) temple-roof-
gesture in [75].   These gestures have properties that are different both from those

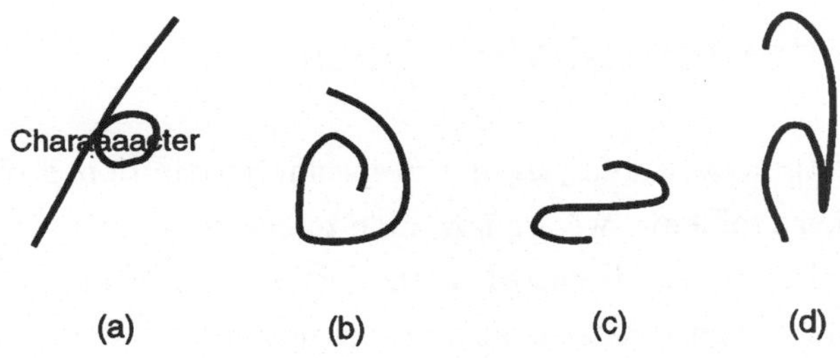

Figure 3.8: Some examples of irregular gestures

of handwritten characters and from handdrawn geometrical symbols. While most
handwritten characters have regular heights and orientations, gestures do not. Further, these gestures differ from regular geometrical figures in that they are freehand
drawings and their features cannot be described with simple functions. For example,
the proofreader's editing gesture used in [73] for delete-gesture (figure 3.8 a) which
is a loop with a beginning and ending tail, can differ in size rotation, and mirror
image.

Compared to character recognition, there are only a small number of investiga-
tions to gesture recognition. To give a feeling about the relationship between efforts
in character recognition and gesture recognition: in the most comprehensive survey
[122] of on-line recognition, there are only 2 papers devoted to gesture recognition
and over 200 papers about character recognition.

Recent research results [73, 107] indicate that trainable gesture recognizers are
successful for the recognition of irregular gestures. These gesture recognizers are
designed only for single-stroke gestures, they do not satisfy our requirements of

multiple-stroke sketching facilities. On the other hand, the low-level recognizer for handsketch-based diagram editing must provide a recognizer which works without training, because trained recognizers are usually person-dedicated and the training takes long time. The same diagram editor for conceptual design can be used by different users. Further, the drawing directions in irregular gestures usually have meanings. In contrast, drawing directions in our regular geometrical figures are invariants of the same figure. Our low-level recognizer must support all drawing styles which are discussed in section 3.1.3 to fulfill our requirements.

3.3 Fundamental Concepts

In the light of the above analysis of our specific recognition problem and the related recognition problems, we are now able to consider our requirements to make own design decisions. As discussed in the last section, character recognizers and irregular gesture recognizers treat related recognition problems with different properties. Character recognition systems and irregular gesture recognizers are pattern classification systems which cannot recognize all graphical attributes. The methods developed for on-line recognition of characters or irregular gestures cannot be used directly in geometry recognition. For this reason, we designed a hierarchical and incremental recognition method. This section describes the fundamental concepts, the major design considerations and design decisions of the low-level recognizer.

3.3.1 Hierarchical Classification

In the requirements discussed in the introduction of this chapter, the response time has been clearly stated as one of the most important criteria for the acceptance of a gesture recognizer for interactive applications. Matching strategy is the most crucial factor which influences the response time. The matching strategy of existing on-line pattern recognition systems normally matches an input pattern with all standard patterns (prototypes) in a dictionary (table) by using some appropriate measurements or calculations. The distance of measurement values between input pattern and standard symbols in the dictionary can be calculated in different manners. Examples of recognition systems which use this strategy are [82, 121, 107]. Therefore, matching is based on different measurements and calculations by minimizing or maximizing evaluation functions. In [82], the Euclidean distance between prototypes and normed patterns is used, in [121] a cumulative distances of angle and height differences are used. Dean Rubine's method maximize a linear evaluation function over 13 feature calculations. The classifier simply determines the class for which the evaluation function gives a maximum value. For a generalized low-level recognizer which must recognize all reasonable handsketched geometrical figures, this matching strategy results in an enormous number of distance calculations which makes immediate response difficult.

Geometrical objects have a clear and intuitive hierarchy. For example, circle could be a subclass of ellipse, or rectangle could be a subclass of parallelogram. This inherent hierarchical nature of geometrical figures can be used to make the classification of handsketched figures more efficient. *At first, some broad classes are distinguished. In the next step each of the broad classes is further subdivided. Subdivision continues until a final classification is obtained.*

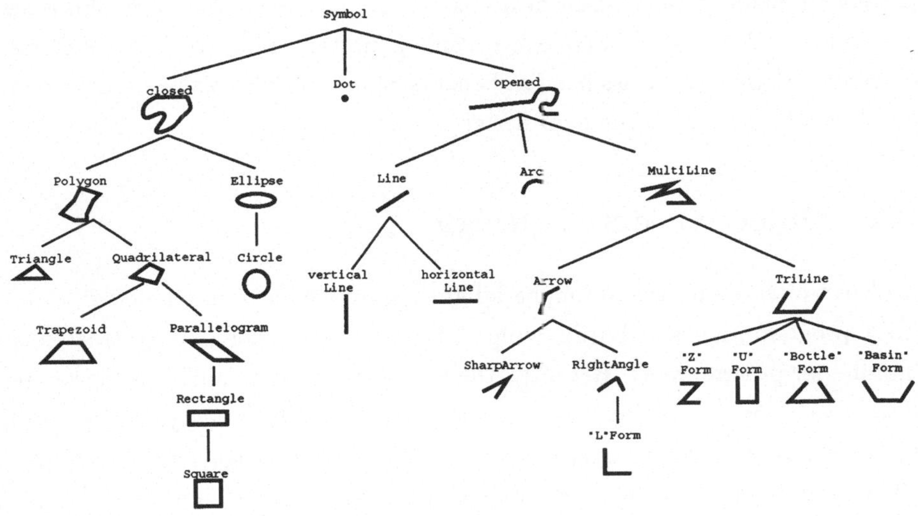

Figure 3.9: The hierarchy of geometrical objects

Figure 3.9 shows the currently used symbol hierarchy. Each new level in this hierarchy is defined by emphasizing the existence of a special feature. For example, if two adjacent edges of a rectangle are equal, this rectangle is a square. In general, all geometrical objects have simple definitions which can be used as matching criteria to classify them.

In figure 3.9, the characteristics of each class are illustrated by giving example-figures below the class-name. For different diagram types, this class-hierarchy can be redefined for the benefit of the recognition task. Experience shows that the system works better if the significant classifications are made at the top level of the hierarchy. For diagrams which mainly distinguish between nodes and connections between nodes, we found that the hierarchy which first distinguishes between opened strokes and closed strokes works best. For other diagrams, the hierarchy can distinguish line-oriented geometrical objects from arc-oriented geometrical objects at the first level of the hierarchy. Although there are many possibilities to define this hierarchy, the recognition method is designed to be independent of a concrete hierarchy. Further, each defined hierarchy can be even adapted or extended for specific applications.

The method of *hierarchical classification* is efficient because only necessary calculations are made for local classifications. A complex recognition problem is separated into several layers, and in one layer, only a limited number of different classes is considered. Hierarchical classification benefits significantly by using object-oriented techniques which leads to the next design consideration.

3.3.2 Object-Oriented Design

The object-oriented programming paradigm has evolved as a methodology to make quality programming less burdensome. It has been predominantly treated as a technique for programmers. Recently, the object-oriented paradigm is viewed more and more as a way of thinking and doing rather than simply as a way of programming [29]. Object-oriented technology is used in the low-level recognizer not only for the implementation of the recognition system, more important, it is used to design the system.

Within our object-oriented design, the classification-hierarchy is used directly as the class-hierarchy. A concrete design of such a hierarchy must satisfy the following two basic requirements:

1. The root class of this hierarchy is the class for original unknown symbols.

2. Each child-class in this class-hierarchy is defined by specifying features which

do not exist in its parent class.

The inherent hierarchical nature of geometry is often used as a standard example to capture features such as inheritance and polymorphism in object-oriented programming as in [41]. The text book example for polymorphism is the definition of a virtual graphical output function *draw* for each geometry class [11]. In this work, the hierarchical nature of geometrical objects is used not for drawing graphics but for recognizing handdrawn graphics to get the geometrical data of the handdrawn graphics. It is shown in [146] that the object-oriented programming is very powerful in applications of on-line recognition of handdrawn graphics. Using object-oriented design, this complicated pattern recognition problem of handsketched geometrical figures achieves an elegant solution. The main object-oriented aspects which strongly influence our design of the low-level recognizer are encapsulation, polymorphism, and reuse.

3.3.2.1 Encapsulation

One of the most significant features of the object-oriented methodology is *encapsulation*, that is, the packaging technique. As discussed in section 3.2, on-line geometry recognition consists of both pattern classification and pattern analysis. We use the encapsulation technique to bring pattern classification and pattern analysis together into objects.

To encapsulate originally unknown strokes and all recognizable geometrical figures into objects of appropriate classes requires the careful design of each individual class in the hierarchy. Each class in the class-hierarchy provides data structures for storing attributes which are relevant for this class.

- The root class *Symbol* is the most general class, and it provides most general data structures for stroke objects which are represented by x-y coordinates.

- All other classes inherit the basic attributes of the root class, and they provide additional data structures for storing class-specific attributes.

In the low-level recognition, input strokes and all recognizable symbols are encapsulated into classes in a class-hierarchy of symbols. The class name of each output object represents the classification result, and the graphical attributes of each object represent the analysis result. The class-hierarchy corresponds to the classification-hierarchy.

3.3.2.2 Polymorphism

A polymorphic function is one that can be applied uniformly to a variety of objects. Class inheritance is closely related to polymorphism. The same operations that apply to instances of a parent class also apply to instances of its subclasses. This property is used to design a uniform classification function for all classes. Each classification function can be accessed by a corresponding object.

- All classes provide a uniform matching-routine which determines whether an object of this class can be an object of its children's class.

- The matching-criteria used for corresponding matching-routines are exactly the definition which distinguishes the child classes from the parent class.

Within this object-oriented design, the recognition process is treated as an *object-refinement* process. Input strokes are encapsulated into objects. Each object recognizes the next possible subclass in the class-hierarchy and creates a new object of that more specific class. The key point here is that the *object recognizes itself* which is supported by the polymorphism concept of the object-oriented technology. All classes in the class-hierarchy specify a virtual recognition function with a unique function name. In this way, the object refinement can be easily controlled by iteratively calling each object to recognize itself [146].

3.3.2.3 Reuse

Reuse of object-oriented technology is represented in the low-level recognizer in two aspects. First, the recognition of multiple-stroke sketches reuse the single-stroke analyzer by merging multiple-stroke symbols into objects which can be considered as a single rough sketch. Second, subclasses in the class-hierarchy inherit graphical attributes and recognition functions of all their superclasses. That is, a subclass reuses the recognition functionality of its superclasses automatically. This has the advantage that a new symbol can easily be extended in the symbol hierarchy by reusing the recognition functionality of its superclasses.

3.3.3 Incremental Recognition

With any user interface, there must be some signal by which the computer knows that the user has completed a command which should now be executed. This signal is called a closure, and it may be an explicit event such as a button press or return-key, or it may be automatically recognized. Rhyne [101] first discussed this *closure-problem* within gestural interfaces. In principle, one would like to avoid the need for explicit closure actions, as they consume time and destroy the directness.

For this reason, only the very early approaches, e.g. [101] use a closure button, most existing gesture-based systems, e.g. [134], determine the closure by using time information. This kind of interaction forces the following style: First, the user draws something. Then he stops drawing for a while to indicate to the system that he has finished drawing the objects. The system waits for a predefined time period during which no coordinates have been sent to the recognizer, a recognition process can then be started. This interaction style is not suitable in diagram editing for conceptual sketching because pauses destroy the thinking process. Using pauses additionally for closure signals, the user feels unpleasant while sketching.

To give the user a better feeling about the directness of the gestural interface, the following design decision is made: On the one hand, we simulate paper and pen by using inking. The user sees what he has just drawn in the same style like working with paper and pen. On the other hand, we improve the paper and pen style by automatic redisplaying each handdrawn stroke with the recognized graphical object, that means, after receiving the pen-up signal, this stroke will be recognized immediately, without the need of pauses. Moreover the recognition result is displayed directly after the stroke is recognized. In this way the user sees immediately the recognized stroke and gets a direct feeling of the underlying gestural interface. The current recognition results are always visible, because the handdrawn stroke is beautified by a regular geometrical object which gives a better display than inking. Inking is only used while the user is drawing. Additional sketchings can be made more precisely, because the previously drawn strokes were already recognized as geometrical objects which are displayed regularly. This has the additional advantage that the user can correct wrong recognitions as early as possible.

The design decision for immediate recognition of each stroke makes it difficult to determine closures of multiple-stroke gestures. This is especially true when two shapes differ in form only by the addition of one or more strokes. The idea for classifying the gesture recognition in two tightly cooperative levels is essential to make it possible that the high-level recognizer helps the low-level recognizer to recognize closures without any user's actions. The communication between low-level and high-level recognizers is supported by a database of geometrical objects. After the low-level recognition process is terminated, the high-level recognizer is triggered to look in the database for syntactically correct geometrical objects. Objects which compose a syntactically correct editing command will be removed from the database. In other words, all geometrical objects, which are stored temporally in the database, are syntactically incorrect. This *incorrectness* is mainly *incomplete*, and incomplete strokes can be successively completed by drawing additional strokes.

This leads to an incremental recognition of multiple-stroke sketches with the following two key points:

1. A database buffers incomplete strokes until they become a syntactical correct sketch.

2. The system incrementally merges all connected and incomplete strokes into new objects. New merged objects are made immediately visible to improve the directness.

The idea of *incremental merging* provides a very powerful mechanism to drastically simplify the recognition efforts for stroke-order-independent multiple-stroke sketches. This is because logically connected strokes are successively merged into one object.

3.4 System Design

3.4.1 Overview

A recognition system is characterized by its control structure and, for this control structure, an appropriate data structure. Figure 3.10 is a schematic of the system components and its organization of the low-level recognizer. The low-level recognizer consists of three modules and a database which is accessible from all modules. The three software modules are the single-stroke analyzer, the incremental updater, and the selective matcher.

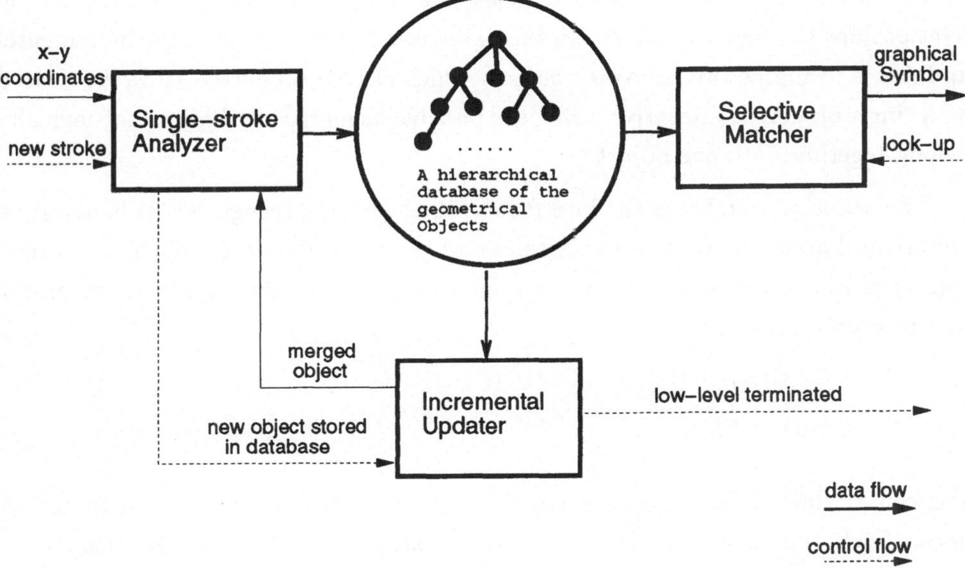

Figure 3.10: System components and organization of low-level recognizer

The intelligence of our low-level recognizer is concentrated in a class hierarchy which corresponds to the symbol hierarchy presented in section 3.3.1. One of the most significant features of our design is that this class hierarchy of geometrical objects is the control structure of the recognition process, and at the same time it builds the storage and query structure of the database. The database is designed for hierarchical access of geometrical objects. Hierarchical access here refers to queries which can be expressed hierarchically, for example, select a geometrical object which belongs to the opened class *or* child-class of the opened class. This kind of hierarchical accesses is used intensively by the incremental updater.

3.4.1.1 System Components

The *single-stroke analyzer* is the interface to the dialog manager. For each new stroke, the single-stroke analyzer creates a new object which encapsulates this stroke which is originally represented by x-y coordinates. This object will be classified along the defined class hierarchy by successively analyzing features of this object.

In contrast to the single-stroke analyzer, the *incremental updater* operates on the relationships between two objects in the database. The main goal of the incremental updater is to merge two arbitrary objects which can be connected and represented by a single object. By iterative calls, it is possible to merge all objects that logically belong together, into one object.

The *selective matcher* is the interface to the high-level recognizer. It is mainly a hierarchical access module to the database of geometrical objects. Each syntactical correct object in the database will be selected and sent to the high-level recognizer by the selective matcher.

3.4.1.2 Control Flow

The control flow of the low-level recognizer is represented in figure 3.10 with dotted lines. Each time a new stroke is completely drawn, the single-stroke analyzer is activated. The termination of the recognition process of the single-stroke analyzer triggers the incremental updater. The incremental updater stops if all objects, which can be combined together, are merged and processed. The termination of the update process triggers the high-level recognizer which will be discussed in the next chapter. The control flow between database and other modules are not depicted in figure 3.10, as they are standard database operations such as storage, delete, and query of objects. The selective matcher can only be activated by the high-level recognizer.

3.4.1.3 Data Flow

The input data of the low-level recognizer are digitized x-y coordinates. Sequences of x-y coordinates are segmented into individual strokes by pen-down and pen-up signals. The outcome of the single-stroke analyzer is an object representing the complete recognition results which include the class and the attributes. The class, this object belongs, is the recognized class of the input stroke, and the graphical attributes of this object are the recognized feature points. This resulting-object will be stored in the central database of geometrical objects. Data exchanges between the database and other modules are based on objects which are instances of classes in a class-hierarchy of geometrical figures. The incremental updater merges every two logically connected objects into one object, and sends the merged object as a single stroke to the single-stroke analyzer. The old objects are deleted after the merging. Data exchanges between low-level recognizer and high-level recognizer are achieved by the selective matcher which takes objects from the database and brings them to the high-level recognizer. The output from the selective matcher are geometrical objects which can be used by the high-level recognizer. The following three sections describes the low-level recognizer in detail.

3.4.2 Symbol Database

One of the most important system components of the low-level recognizer is the central database. This database is used for storing intermediate recognition results which are shared by all the three modules: the single-stroke analyzer, the incremental updater, and the selective matcher. As mentioned above, all intermediate recognition results are objects of classes in our class-hierarchy. These objects have a hierarchical *kind-of* relationship among them. For example, a square-object is a kind of rectangle-object. This hierarchical relationship is important for the incremental updater as well as for the selective matcher. The incremental updater has to check connectivities between objects which belong to the class **opened** and all its subclasses. The selective matcher accesses objects in the same way. The key point here is the so-called hierarchical query, that is, searching for objects which belong to a specific class *or* objects which belong to the child-classes of this specific class. This can be seen more clearly in a concrete example. In a Petri net editor, a rectangle is a gesture for "create transitions." If a square is drawn, a transition object must be created, because a square is a kind of rectangle.

A database, which allows hierarchical retrieval of objects, is designed by using the class-hierarchy as the data organization structure in the database. A container class is designed to manage instances of classes in the class-hierarchy. The low-level recognizer automatically creates a container-object for each class in the class-hierarchy. These container-objects are then connected in the same structure as the class-hierarchy shown in figure 3.11. Instances of the same class are stored in a list which is accessible from the corresponding container-object. Additional information such as the number of all children objects which are instances of its subclasses

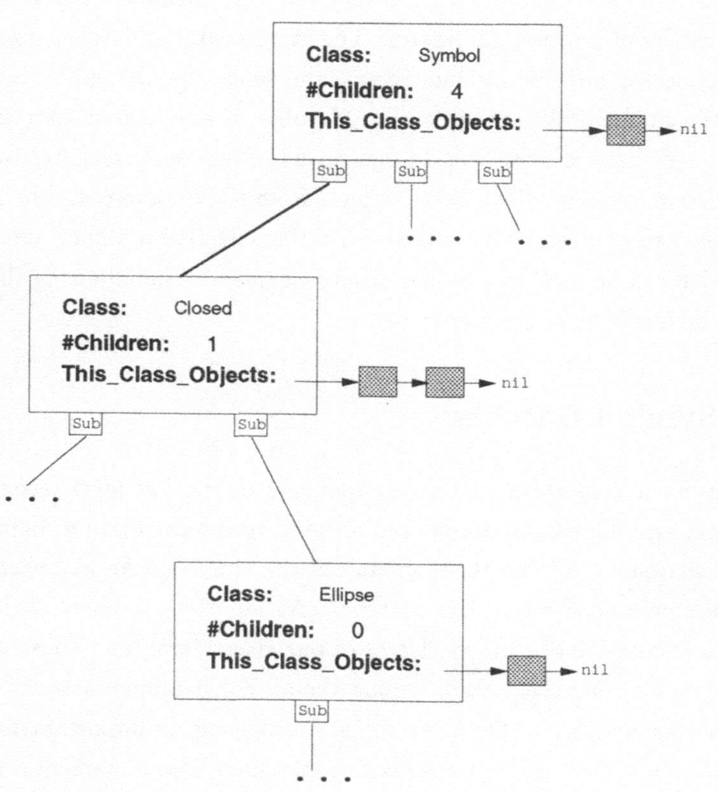

Figure 3.11: Internal structure of the specific database for geometrical objects

are stored in each container-object. A hierarchical query is therefore a top-down-oriented search from an entry container-object by looking for a non-empty list of object-instances. In this way, a hierarchical query of objects stored in the database is achieved.

3.4.3　Single-Stroke Analyzer

The single-stroke analyzer transforms a single handsketched stroke into a geometrical object. Handsketched figure recognition requires pattern classification and pattern analysis. One of the novelties of our single-stroke analyzer is that classification and analysis are combined in a natural way by using an object-oriented method.

3.4.3.1 Control Structure

Single-stroke analysis may be viewed as a problem-solving activity which comprises both pattern classification and pattern analysis. The initial state of the problem is defined by the original stroke. By a sequence of actions, the initial state undergoes a sequence of state transitions which leads to a sequence of new states. The single-stroke analyzer stops if no further action is possible, and the "stop-state" is the goal state.

One significant characteristic of the low-level recognizer is that the control structure of the single-stroke analyzer is exactly the class-hierarchy. The class-hierarchy can be seen as a schema for a top-down problem-solving strategy, as well as a knowledge base which reduces the searching space in the problem-solving activity. The root class represents the initial state, and all other classes represent goal states. This differs from conventional problem-solving-trees such as decision trees which consider only leafs as goal states [89].

As mentioned above, each class in the class-hierarchy provides data structures for encapsulating strokes and geometrical objects, and each class has an analyze-function for local recognition. *Local* means that only one level of the hierarchical classification is considered. Within object-oriented design by making use of the polymorphism, the control structure of the single stroke recognition is just a simple loop as shown in figure 3.12.

Firstly, each new stroke is encapsulated in an object of the root class **Symbol**. We call this object the working-object. Subsequently, the recognition process carries on by calling the uniform and polymorphic analyze-function for this working-object in a loop. Each analyze-function returns a new object which represents the local recognition result of the working-object. Within the loop, the new object is assigned to the working-object in case that they are different, that is, the working-object is specialized. In this way, the working-object is recognized step by step. The class and the attributes of this working-object changes top-down along the class-hierarchy. The recognition result is represented by the working-object after the termination in case that the working-object cannot be specialized any more.

The essential point here is that all classes in the class-hierarchy have a uniform analyze-function. This polymorphic analyze-function is a local decision maker which determines whether an object of this class can also be an object of a more specific

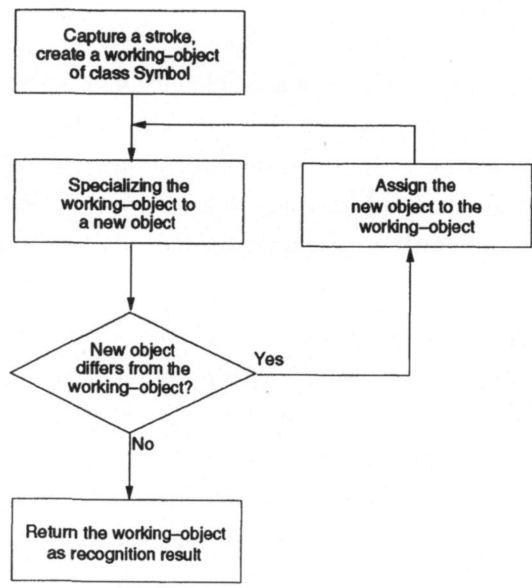

Figure 3.12: The control structure of the single-stroke analyzer

class, that is, whether this object can be further "specialized". All analyze-functions comprise both pattern classification and pattern analysis. The classification is characterized by the difference between the class-name of the object itself and the class-name of the returned object which represents the recognition result. The pattern analysis is characterized by the transformation of object-attributes.

Because this function is a polymorphic function, it can be used for all objects which are instances of different classes in the class-hierarchy. In the recognition loop, each working-object calls this function for itself, and the returned object of this function is assigned to this working-object again.

An Example

The best way to illustrate the recognition process is by example. For this reason, we consider the handsketched rectangle which was discussed in section 3.1(figure 3.1) as an input for our single-stroke analyzer. Figure 3.13 illustrates the recognition process by giving the intermediate steps which are represented by the working-objects created during the recognition process.

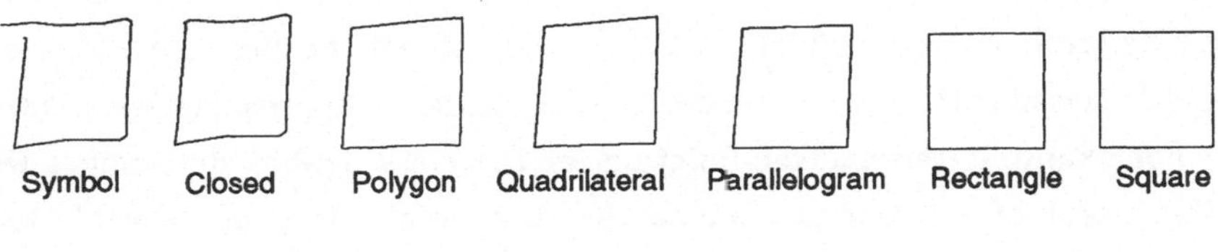

Figure 3.13: The recognition process of a single-stroke square

First, the original stroke is encapsulated in an object of the root class `Symbol`. The analyze-function of class `Symbol` finds out that this object belongs to class `Closed`, and an instance of the class `Closed` is then created and returned. This step includes both pattern classification and pattern analysis. The working-object is classified from a general class `Symbol` to the more specific class `Closed`. Moreover, the specification data of the working-object is analyzed. The start point and the stop point of the working-object is characterized in the new `Closed`-object by a single point.

This new object is then assigned to the working-object for further recognition. The recognition process carries on with this new working-object. The working-object is subsequently recognized as a `Polygon`-object, and then a `Quadrilateral`-object, a `Parallelogram`-object, a `Rectangle`-object, and finally a `Square`-object which represents the recognition result.

3.4.3.2 Feature Selection and Feature Analysis

Given the control structure, the design of the single-stroke analyzer is the design of the analyze-functions for each class. These functions are all based on feature selection and feature analysis. One of the advantages of hierarchical classification is that a complex recognition problem is separated into many simple recognition problems, so that most of the analyze-functions are easy to design. For example, the analyze-function for the class `Polygon` considers the number of vertices of a `Polygon`-object. A polygon with three vertices is a triangle, and a polygon with four vertices is a quadrilateral. Other functions consider quantitative properties of an object such as the slope angle of a line or distances between feature points.

Various thresholds are used for making decisions. These threshold-values are defined dynamically, dependent on some other object-specific properties, so that reasonable classification can be made. This is similar to approaches with fuzzy logic [133]. For example, the analyze-function for the class Symbol determines whether the start point of a stroke is close to the stop point. It is not feasible to use a fixed threshold to determine, if the distance between the two points is less than this value it is a closed object, otherwise it is an opened object. The closeness here is vagueness and ambiguity. Consider a big circle-stroke with a radius of about 100 pixels, if the distance between the start point and the stop point of this stroke is 20 pixels, it is reasonable to classify this stroke as a closed stroke. But if the same distance is considered for a small circle-stroke, for example, a radius of about 30 pixels, the stroke should be classified as an opened stroke.

Additional to these forms of simple feature analyses improved by the use of fuzzy logic, corner detector, line detector, and arc detector are more difficult, which need further discussions. First, we consider the representation forms of a stroke object.

Stroke Representation Forms

Each coordinate pair defines a point p_i with its x- and y-coordinate:

$$p_i := (x_i, y_i)$$

The sequence of these coordinate pairs builds a list P of coordinate pairs, which is the original input for further recognition tasks:

$$P := \{p_i \mid 1 \leq i \leq n\}$$

This list of coordinates is used as object-attributes of the root class Symbol, all other classes in the class-hierarchy inherit this list as the basic representation form of all stroke objects.

Each point (except the last one) together with its next point defines an angle a_i which represents the direction of the drawing at that point.

$$a_i := \begin{cases} arctan\left(\frac{y_{i+1}-y_i}{x_{i+1}-x_i}\right), & \text{if } y_{i+1} \geq y_i \text{ and } x_{i+1} \geq x_i \\ 360 + arctan\left(\frac{y_{i+1}-y_i}{x_{i+1}-x_i}\right), & \text{if } y_{i+1} < y_i \text{ and } x_{i+1} \geq x_i \\ 180 + arctan\left(\frac{y_{i+1}-y_i}{x_{i+1}-x_i}\right), & \text{if } y_{i+1} \geq y_i \text{ and } x_{i+1} \leq x_i \\ 180 - arctan\left(\frac{y_{i+1}-y_i}{x_{i+1}-x_i}\right), & \text{if } y_{i+1} < y_i \text{ and } x_{i+1} \leq x_i \end{cases}$$

All angle values build the list A of angles:

$$A := \{a_i \mid 1 \leq i < n\}$$

A widely used encoding method for input coordinates is Freeman's chain code [32], which is designed originally as a more memory-saving representation of point-pictures. In the single-stroke analyzer, this coding method is used to represent the directional information.

$$f_i := (a_i \ div \ 45) \ modulo \ 8$$

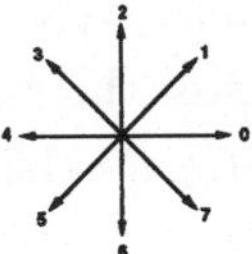

Figure 3.14: The eight possible directions of the chain-code

The list of f_i build the chain code representation of a stroke:

$$F := \{f_i \mid 1 \leq i < n\}$$

The list F of chain codes can be seen as a filter of the information represented in the list of angle values.

Corner Detector

Corners are important features used by the single-stroke analyzer to classify closed strokes to polygons, and to classify opened strokes to multilines, respectively. Corner detection techniques have been widely used in many applications involving shape analysis. Liu and Srinath [74] have evaluated a number of boundary-based corner detectors. The basic principle of such corner detectors is the following: If, at a point, the object boundary makes discontinuous changes in direction, or the curvature of the boundary is above some threshold, then that point is declared as a corner point. The Rosenfeld-Johnston corner detector [103] is found most appropriate as a basis for detecting corners in handsketched strokes.

A k-vectors at point $p_i = (x_i, y_i)$ is defined as:

$$\mathbf{a}_{ik} = (x_i - x_{i+k}, y_i - y_{i+k})$$
$$\mathbf{b}_{ik} = (x_i - x_{i-k}, y_i - y_{i-k})$$

the k-cosine, which is the cosine of the angle between $\mathbf{a}_{ik}$ and $\mathbf{b}_{ik}$, as follows:

$$c_{ik} = \frac{\mathbf{a}_{ik} \bullet \mathbf{b}_{ik}}{|\mathbf{a}_{ik}||\mathbf{b}_{ik}|}$$

An appropriate value of k at each point i is selected as follows: First, using the k-cosine definition to compute the values $c_{i1}, c_{i2}, ..., c_{im}$, whereby m is chosen dependent to the number n of coordinates of the stroke-object. The best values of k is chosen such that the following conditions holds:

$$c_{im} < c_{i,m-1} < ... < c_{ik} \geq c_{i,k-1}$$

The value of k and the corresponding c_{ik} are used to detect corners. The point p_i is a corner if c_{ik} is a local maximum for all j such that $|i - j| \leq k/2$.

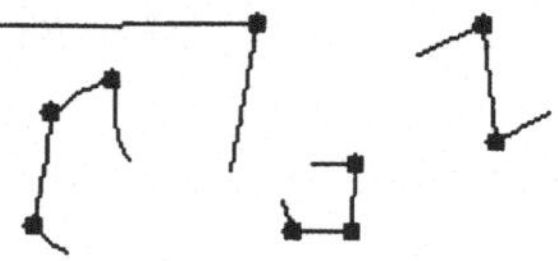

Figure 3.15: Detected corners are marked by black dots.

This and all other existing corner detectors are designed mainly for off-line recognition of closed shapes which are represented by chain code after segmentation. Improvements and modifications are needed for using them in on-line recognition. One problem caused by this corner detector for handsketched strokes is that it usually finds many corners which are not corners in human's perception. This is because handsketched strokes often contains noise due to the characteristics of handdrawing. As a solution to this problem, a filter is designed, that makes use of chain code [32] as a post processor for the corner detector. The basic idea is to eliminate corners where the chain codes near this corner have equal directions. Figure 3.15 shows the results of the improved corner detector used on some single strokes.

A Rosenfeld-Johnston corner detector is appropriate for objects which have relatively many points. As stated above, the single-stroke analyzer is also used to classify merged objects which are outputs of the incremental updater. These objects usually have only a few points. For example, a line merges with another line to build an arrow-object represented by only 3 points. To find corners within such objects, a Rosenfeld-Johnston corner detector is not feasible, because the k-cosines cannot be calculated reasonably in this case. For this reason, a chain-code-based corner detector is designed as a supplementary method for finding corners in objects with only a few points. The basic algorithm is to consider points as corners where the chain code changes its value. Both methods are encapsulated in a corner-detector class within the meaning of object-oriented programming. Dependent on the number of coordinates, an appropriate corner detector can be chosen automatically.

Line Detector

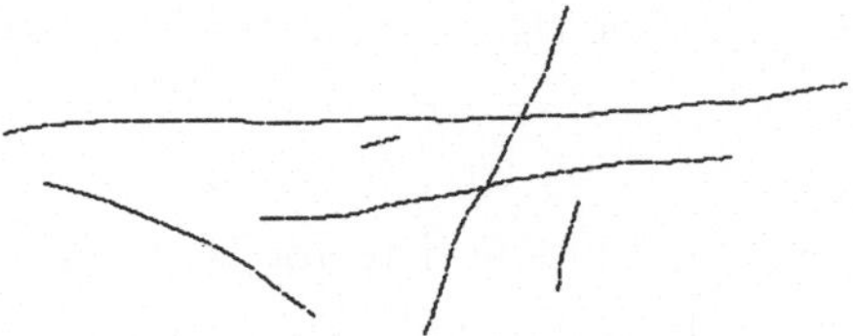

Figure 3.16: Some typical handdrawn lines

A line detector is used to examine whether a sequence of point coordinates forms a line. Usually, there is much more noise in handsketched lines than in lines which are considered by image recognition. Therefore, it is not easy to check whether a stroke or a part of a stroke is a line or not. Figure 3.16 shows several handsketches which are recognized by the single-stroke analyzer as lines. One of the characteristics

of handsketched lines is that there are many "waves" in such lines. This is because
the movement of free hand drawing cannot be controlled to keep straight.

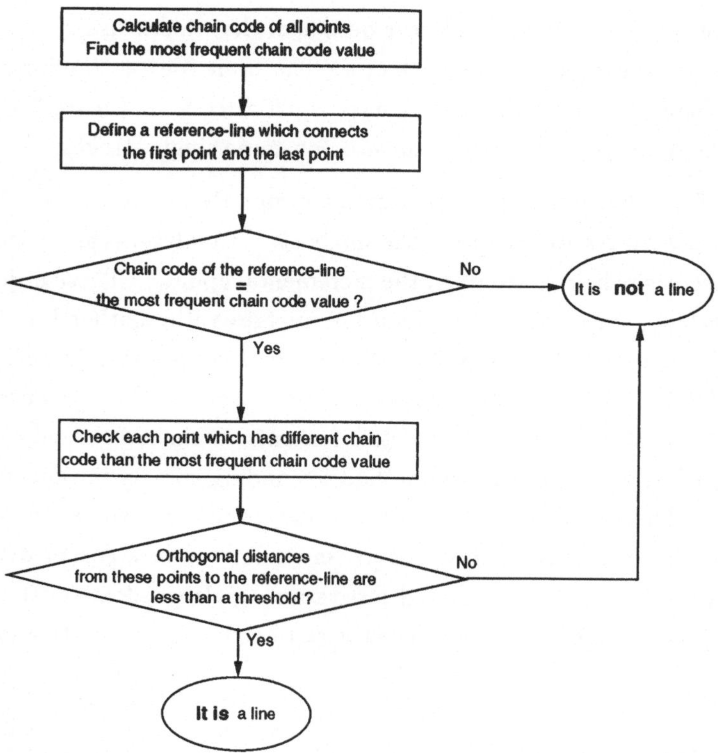

Figure 3.17: The algorithm of the line detector

The algorithm designed for the line detector is shown in figure 3.17. Firstly,
the test-object is assumed to be a line, and the chain code values of all points are
calculated. The most frequently used direction in the chain codes is defined as the
direction of the assumed line. Secondly, a reference-line is defined by using the first
and the last coordinate points of the test-object. Then, a primary examination is
done by comparing the direction of the reference-line and the test-object. If the
chain code value of the reference-line is different from the chain code value of the
assumed line, the test-object is not a line. If the directions are the same, additional
tests are done by comparing the chain code values at each point with the chain code

of the assumed line. At a point, whose chain code differs from the assumed line direction, the distance from that point to the reference line is calculated. If all these distances are less than a threshold, the test-object is then a recognized line object. To improve the line recognition, a dynamical threshold is used, which depends on the length of the reference line.

Arc Detector

An arc is considered to be a handdrawn ellipse, a circle, or a part of them. These include general conics such as ellipses with tilted axes, hyperbolas, and parabolas which can be described as:

$$Ax^2 + Bxy + Cy^2 + Dx + Ey + F = 0$$

Figure 3.18 shows some examples of arcs which can be recognized by the single-stroke analyzer. Similar to handsketched lines, handsketched arcs usually have "noise".

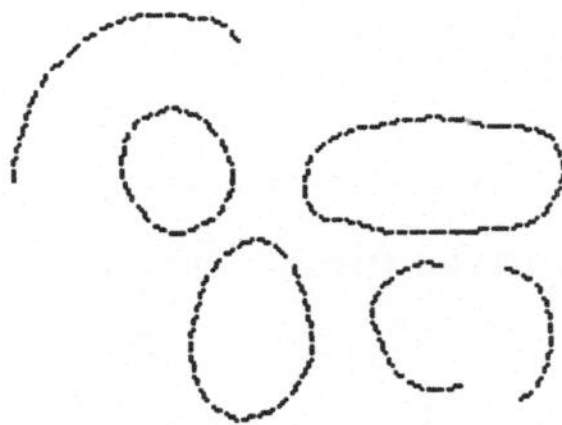

Figure 3.18: Typical handdrawn arcs

The basic idea of the arc detector designed for the single-stroke analyzer is based on the above mathematical specification of general conics. It is a well-known fact that there are no inflexion points in functions of general conics. Analyzing the handsketching process, it is obvious that an inflexion point arises only if the drawing direction changes. Under this consideration, a very efficient and effective algorithm is designed to check whether a stroke-object is an arc or not. Eliminating the implementation details, the main points of the algorithm are described below:

1. calculate the chain code values for each point.

2. find out the drawing direction by using the first three chain code values which are not equal, an arc has two possible drawing directions: clockwise or counterclockwise.

3. check whether all other points have the same drawing directions, local noise can be minimized by look ahead techniques.

While the classification of arcs is easy, a general analysis of the exact arc-parameters is not trivial, particularly the analysis of parameters of an ellipses with tilted axes. This can be done by solving a nonlinear equation system to find out the five coefficients A, B, C, D, E, F in the general conics equation, but this method is very inefficient and slow. Because only simple geometrical figures are used in diagram languages, ellipses whose axes are tilted, are not considered in this work. Therefore, all arcs can be approximated with ellipses or circles whose axes are aligned with the axes of the plan. Such arcs can be specified simply by giving the bounding box.

3.4.4 Incremental Updater

Diagram symbols can be drawn both with a single stroke and with several strokes one after the other. Usually humans prefer the successive drawing style which generates the so-called multiple-stroke sketches. On-line recognition of multiple-stroke sketches is difficult. In section 3.1.3, we have discussed the properties of multiple-stroke sketches. One of the problems of recognizing multiple-stroke sketches is the treatment of numerous possibilities of stroke combinations. Another problem is the automatic closure-determination without any explicit command from the user, because the need to indicate the end of a gesture makes the user interface more awkward than it need to be.

In section 3.3.3, the basic concepts of the incremental recognition method were presented. In order to implement this incremental recognition, a module is needed which incrementally merges connected objects together into new objects. This module is called the incremental updater. The main task of the incremental updater is to merge connected objects within the database into new objects. In this way, all connected objects, which belong logically together, can be merged into a single object.

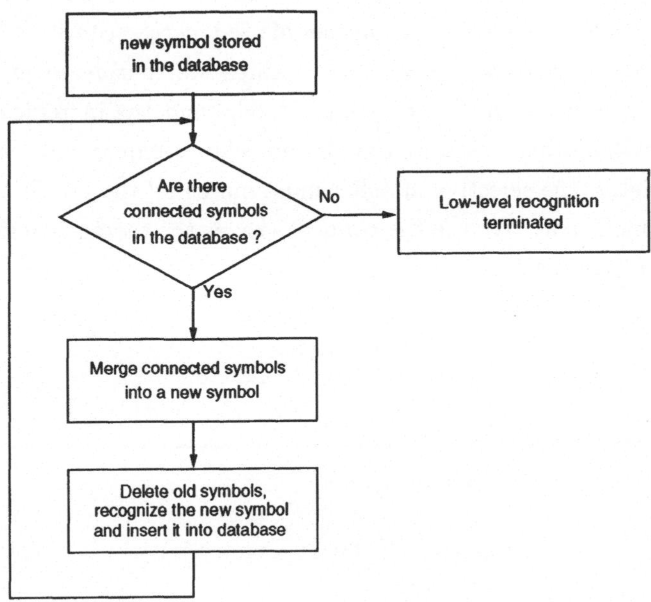

Figure 3.19: The incremental updater merges all connected symbols in the database into a new symbol.

Figure 3.19 illustrates the control structure of the incremental updater. After a new stroke is recognized and stored in the database by the single-stroke analyzer, the incremental updater begins to search for connected symbols in the database. In case there are no connected symbols, the updater terminates. In case that connected symbols are found, each two symbols will be merged into a new symbol representing the two in one. Then, the updater removes the connected symbols from the database, recognizes the merged symbol, stores it into the database, and begins to search connected symbols again. In the following, several important aspects of the incremental updater are described.

Criteria for Merging

In contrast to the single-stroke analyzer, the incremental updater operates on the relationships between two stroke-objects. The relationship, which is currently used as the merge-criterion, is the connectivity between two opened objects. In an early version [145], other merge-criteria such as close-to were also used. It has been shown that there is a trade-off between the number of used merge-criteria and the matching task of the selective matcher. If all relationships are considered in the incremental updater, the selective matcher becomes trivial and the incremental updater is overloaded. In the other extreme, that is, no relationship is considered in the incremental updater, the selective matcher must match all the possible combinations which is very inefficient. We use the connectivity as the merge-criteria, which gives the best recognition results.

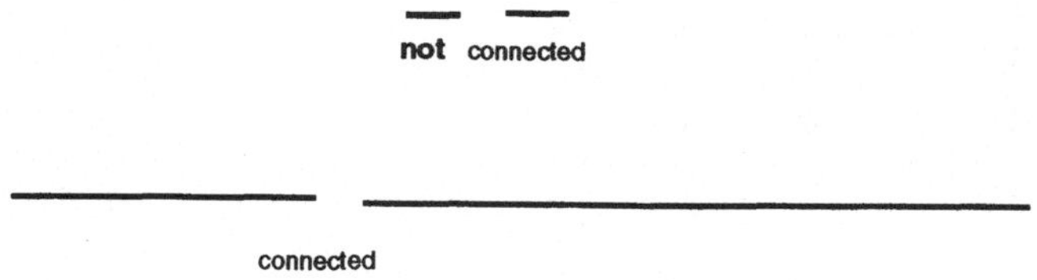

Figure 3.20: Fuzzy Connectivity

Fuzzy Connectivity

Fuzzy logic is used for the detection of the connectivity between two objects. The reason is that the connectivity cannot be computed absolutely by using a fixed threshold. For example, to check whether two lines are connected to each other, the dimension of these two lines must be considered. Two lines with a length of about 600 pixels can be seen as connected to each other, if the distance between their endpoints is smaller than 60 pixels. However, if the lines are only 50 pixels long, which is even smaller than the distance between their endpoints, they are obviously not connected. For considering such problems, fuzzy logic is the best solution. The connectivity is therefore calculated under considerations of size and dimension of the examined objects. We use a dynamical threshold which is dependent on the sizes and other graphical attributes of the examined objects. Within object-oriented programming, all necessary parameters are encapsulated into objects, the dynamical threshold can be calculated easily.

Heuristics

The searching process for two objects which are connected logically, can be optimized by using some heuristics. Two heuristics are used in the current implementation. First, the connectivity relationship can only exist between two **opened** objects. The hierarchical database designed in section 3.4.2 allows convenient access to objects which have the **opened** characteristic because all objects are stored hierarchically in the database. For checking connectivity, only objects stored below the node of class **opened**, need be considered. In this way, the number of treated objects is reduced by half. Second, in multiple-stroke sketching, the user usually draws connected strokes directly one after the other. For this reason, the last two objects finally stored in the database, are always checked first.

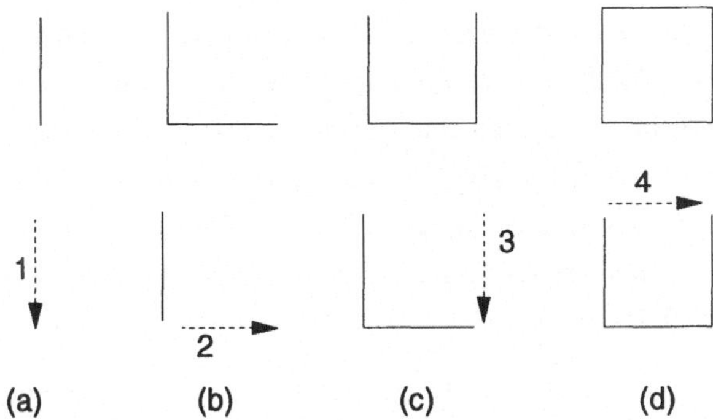

Figure 3.21: Incremental recognition of a multiple-stroke rectangle

Merging Effect

One of the advantages of the incremental merging is that the user can see immediately what happens in the recognition system. Figure 3.21 depicts the merging effect by giving the intermediate steps in a drawing example of a 4-stroke rectangle. In the bottom area of this figure, inkings are drawn by using dotted arrow-lines, and already recognized single-strokes are drawn with solid lines without arrows. In the top area of the figure, the current recognition results are illustrated, these objects are stored in the database. From (a) to (b) the two single strokes 1 and 2 are merged together to a "L"-form (top area of (b)). In the same style, this "L"-form is merged with the third stroke into a "U"-form. As a result, the user sees immediately the effect how the stroke just drawn is composed with other strokes into a new object. He can then draw other necessary strokes incrementally to complete a handsketched geometrical figure into a syntactically correct object.

Reuse of the Single-Stroke Analyzer

If two connected objects in the database are found, a new object will be created by merging these two objects. Merging means that the two connected objects are represented by a new object which must be recognized as well. The question now is how to recognize this object. The first approach was to recognize this object in the routine where this object is created. For example, if two line-objects are merged to a new object, one can test whether the new object is a line or an arrow. During prototyping, the author found that much of the algorithms designed for the single-stroke analyzer can be reused for this purpose because the recognition of a merged object is nothing else than the single-stroke recognition. The single-stroke analyzer can be reused for recognition of merged objects. Therefore, the new merged object can be sent to the single-stroke analyzer in the same way as a new drawn stroke as shown in figure 3.10.

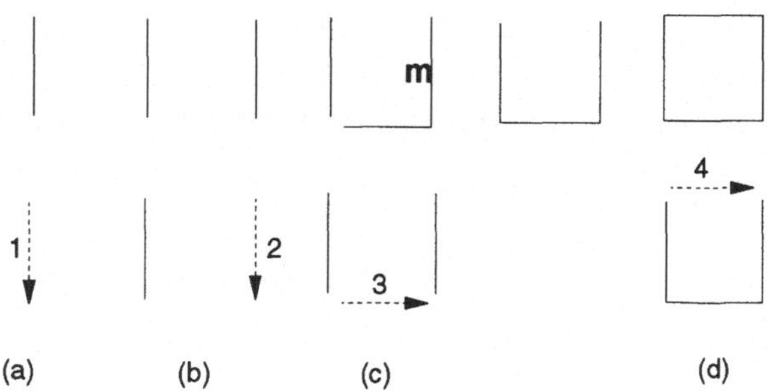

Figure 3.22: An example of iterative merging

Iterative Merging

The incremental updater is an iterative process, that is, if two objects are merged together into a new object, the incremental updater calls itself again to consider this new configuration. To illustrate the effect and necessity of the iterative merging, we consider another possibility for drawing a four-stroke rectangle. The only difference between figure 3.21 and figure 3.22 is the stroke-order. In figure 3.22, the second stroke is not connected to the first one, therefore the incremental updater has no merge-actions. After the third stroke is drawn, the incremental updater finds first that the third stroke is connected with the second stroke, these two objects are then merged to a new object m which will be recognized as a "L"-form object. Because a new object m is created, the incremental updater calls itself again. In the second step, the incremental updater found that the object m is connected to the first stroke-object, these two objects are then merged to an "U"-form object. The last stroke produces the same update-action as that in figure 3.21.

3.4.5 Selective Matcher

One novelty of the low-level recognizer is a *selective* matching strategy which differs from most other existing on-line pattern recognizers. We do not try to group all of the already drawn strokes in a group as a pattern and to calculate the distance between this input pattern and all standard dictionary patterns to get the recognition result by optimizing certain distance calculations such as used in [82, 91]. In contrast, a selective matcher is designed to work together with the high-level recognizer. In other words, the selective matcher is the interface between the low-level recognizer and the high-level recognizer. The high-level recognizer controls the low-level recognizer by calling the selective matcher to get objects from the database.

Each time a new stroke is drawn, the high-level recognizer selects objects in the database via the selective matcher to get syntactically correct sketches. Therefore, the matching process is a selection process which depends on the characteristics of each concrete gesture.

Simple Selection

Gestures for editing diagrams can be classified into two groups. The first group consists of simple geometrical figures which can be arranged to one of the classes in the symbol hierarchy. (see page 49). Gestures in this group are figures which can be drawn both in single-stroke and in multiple-stroke. Independent to how they are drawn, such gestures are finally always single objects stored in the database as a result of the incremental merging. In this case, the task of the selective matcher is just a database access to get such an object, that is, only selection is necessary.

Composite Selection

The second group deals with gestures which cannot be drawn in a single-stroke. As discussed in the previous section, the incremental updater uses only the connectivity as the merge-criterion. This implicates that only figures, which can theoretically be drawn in a single stroke, can incrementally be merged into single objects. Therefore, gestures in this group consist of at least two objects in the database. In this case, the task of the selective matcher is to select possible objects in the database and check the relationships between them. For example, the cross symbol is used as the delete-gesture. This symbol consists of two lines which cannot be drawn in a single stroke. To recognize this symbol, the selective matcher must check all drawn lines to examine whether two lines among them build a cross symbol.

A transistor symbol consists of a circle and four lines which cannot be drawn in a single stroke. Therefore, to recognize a transistor symbol, the selective matcher must select a circle object and four line objects in the database. The graphical attributes of these five objects are examined whether they build a valid transistor symbol. Several template matching methods have been developed to recognize characters or symbols. Some techniques like inter-stroke distances[91] or graph-searching [82] can be used in the selective matcher for this purpose.

As stated in the introduction, the main goal of the low-level recognizer is to recognize simple geometrical figures used in various diagrams, and to allow the user to sketch diagrams in the paper-like style. These symbols are mainly simple geometrical figures. Although complex symbols such as a transistor symbol can be recognized by the selective matcher as well, one should avoid the use of complex symbols as gestures. For example, one can use a triangle as the gesture for "create transistors" in a schematic editor. The design of gestures will be the topic of the next chapter. Nevertheless, to examine the versatility of the low-level recognizer, the author has tried to recognize several complex symbols such as the aforementioned transistor symbol and gate symbols in the current system architecture. The result shows that the object-oriented system architecture can be seen as a framework to integrate many different recognizers.

3.5 Summary

We began this chapter by analyzing the basic requirements, the specific properties, and the key problems of the low-level recognition. We found that on-line handsketch recognition is neither the same as the well-known off-line recognition nor the same as the on-line character and irregular gesture recognition. A novel concept of hierarchical and incremental recognition is presented based on an object-oriented design aimed at handsketch-based diagram editing.

Our low-level recognizer fulfills the requirements for on-line geometry recognition for handsketch-based diagram editing. The hierarchical recognition makes the single-stroke analyzer efficient and robust. The object-oriented technology such as encapsulation and polymorphism allows an easy integration of pattern classification and pattern analysis. Our low-level recognizer is *incremental* and *interactive*, that is, the user immediately sees the recognition result after a stroke is drawn. The recognizer acts as a beautifier which transforms each handsketched stroke into a regular geometrical figure. Incremental merging of logically connected symbols solves the problem of multiple-stroke gesture recognition.

The hierarchical recognition and the hierarchical management of recognized objects in a hierarchical database support the robust and tolerant recognition. Because the hierarchy of geometry is a specialization hierarchy, the hierarchy level represents the *exactitude* of the recognition. Tolerant and robust recognition is possible by proper reduction of this exactitude of the gesture specification which will be discussed in the next chapter.

Chapter 4

High-Level Recognition

A high-level recognizer transforms graphical symbols recognized by the low-level recognizer into commands for creating and manipulating diagrams. In contrast to the low-level recognition, the high-level recognition is editor-dedicated and language-oriented. Therefore, we first introduce the diagram class to which our high-level recognizer is designed, and provide a few formal definitions for handsketch-based editing. Then we concentrate on the fundamental concepts of the high-level recognition, and describe the system components with a number of representative examples.

4.1 Formal Basis

In this section, we formally define the syntactical elements of a very frequently used diagram class by describing the visual alphabets and the visual compositions. We establish an appropriate graph model for the internal representation, and consider both representation forms simultaneously in the definition of *diagram schema*. Then, we introduce the concept of gesture operator and handsketch-based editing.

4.1.1 HiNet Diagrams

In general, diagrams refer to all two-dimensional representations of any kind of information. There are countless diagrams which are used in everyday's life and in science. For example, diagrams for division or multiplication, coordinate diagrams for mathematical functions, and "cake"-diagrams for comparing percentages as shown in figure 4.1.

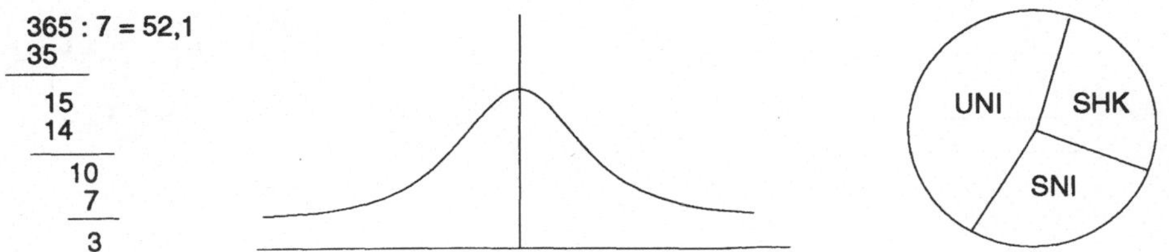

Figure 4.1: Examples of general diagrams

The meaning of diagrams in this work is more specific than this, they are considered as visual programs of certain diagram languages. Diagram languages are the most frequently used visual languages in computer aided design, software engineering, or structural analysis. They play an essential part in designing complex systems and developing programs. Several decades ago, flowcharts were used to design and illustrate programs. Nowadays we have SADT [78], Nassi-Shneiderman diagrams [85], state transition diagrams, statecharts [43], Petri nets [98], Express [114], SpecChart [126], entity relationship charts, semantic nets, influence diagrams [123] and many others. Figure 4.2 illustrate several different diagrams from this category.

The enthusiasm of development and application of so many diagram languages due to the following significant advantages of diagram languages:

- Diagrams give an aid to clear thinking and enforce good structuring. Relationships between objects can be represented much clearly in a diagram than in a textual program.

- Diagrams allow quick and precise communication between members of a development team, because human perception is naturally two or three dimensional rather than one dimensional [104].

These diagram languages are widely used for purposes like analysis, specification, design, modeling, or description of various structures. The syntactic structures of

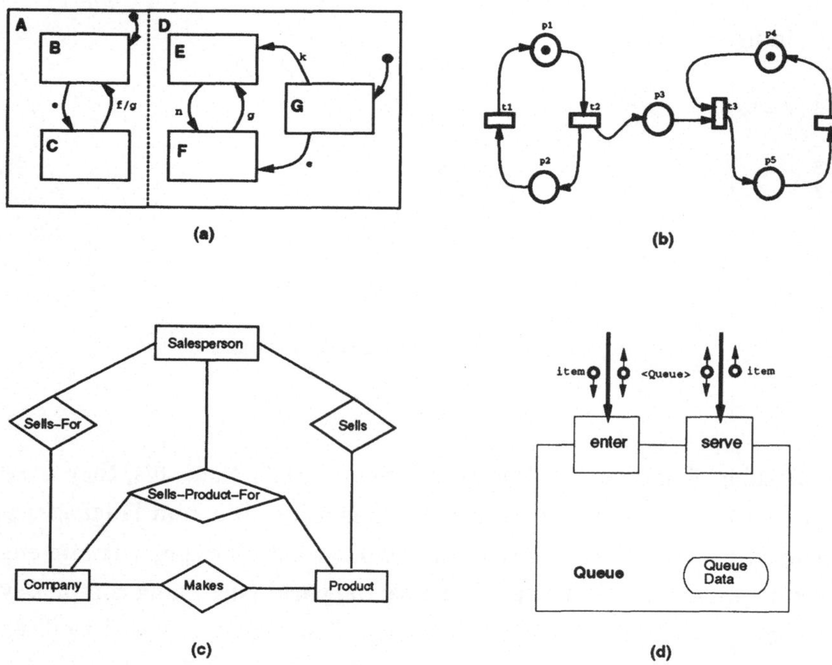

Figure 4.2: a) A statechart, b) A Petri net, c) An entity-relationship chart, d) An OOSD diagram.

these diagram languages have in common that they are of a discrete nature. The essential generality of such diagram languages is clearly stated by Szwillus [120]:

> "They are built up from well-identifiable interlinked graphical objects and absolute size and position of the objects are of minor importance. The main purpose of the pictures is to express *connectivity* and *hierarchy* information."

For the high-level recognition, we abstract this kind of diagram languages to a class called **HiNet** diagrams. **Hi** indicates the *hierarchical* features, and **Net** indicates the reticulate feature of such diagrams.

4.1.1.1 Graphical Structures

A visual language is characterized by the visual symbols and by how the symbols
are arranged to form a picture. The syntax of a visual language is a set of rules
that tell whether a picture is a valid program or not. Therefore, a visual language
specification defines the set of symbols and how these symbols are combined to form
valid pictures.

<u>Definition 4.1</u>

> A *picture element* is a pair, $e = (c, A)$, where c is the *symbol class* of e,
> and A is a finite set of *characteristic attributes*.

□

The *symbol class* specifies the type of the picture element. There are globally
two basic symbol classes used in diagrams as shown in figure 4.3: graphical symbols
such as rectangles, and text fragments such as names or other textual information
attached to a graphical symbol.

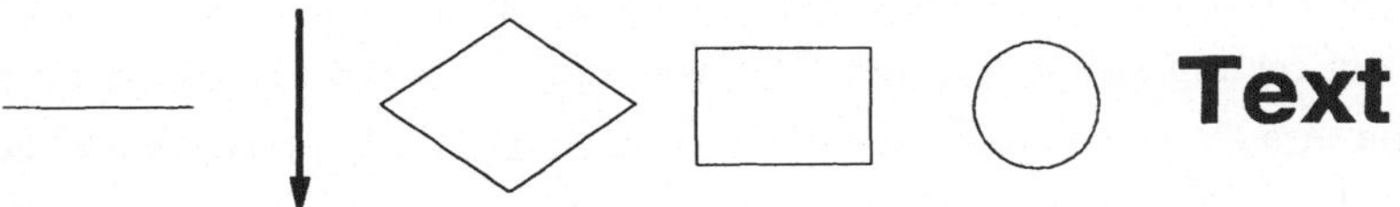

Figure 4.3: Examples of picture elements used in diagrams

The *characteristic attributes* contain all the relevant information about the graphical representation of a specific picture element. The characteristic attributes describe the properties that distinguish picture elements of the same symbol class.
Frequently used characteristic attributes are:

- coordinate values of feature points,

- the character string of a text fragment,

- color, fill pattern, line width, line style.

Definition 4.2

A specific diagram language uses a finite set of symbol classes. This set is defined as the *visual alphabet* of this diagram language denoted by Π.

$\square$

For example, the Petri net diagram language has the visual alphabet $\Pi = \{rectangle, circle, arrowline\}$.

Definition 4.3

A *picture* is a set of picture elements.

$$P = \{(c_i, A_i) \mid (c_i, A_i) \text{ is a picture element.}\}$$

$\square$

This definition is analogous to the definition of a string for a textual language. Object-oriented graphical editors such as `MacDraw`, `GemDraw`, or `idraw`, are based on this picture model. The graphical objects which can be created by these editors are instances of certain *symbol classes*. The user can define the *characteristic attributes* of each picture element interactively by direct manipulation techniques like rubber-banding. The role of this kind of graphical editors is the same as text editors like `vi` or `emacs`.

Pictures of diagram languages are "more" than this picture model defines. This is similar to that a program text is not just a string of characters. The deeper structure of the picture is implicit in the attributes of the picture elements, just as the deeper structure of a string is implicit in the position of the symbols [40]. The relationships between picture elements have a defined meaning. To examine the two-dimensional relationships, we first analyze the compositions of picture elements.

Compositions

In a textual language, concatenation is the only composition which combines adjacent elements in a string. In this manner, substrings are combined to form expressions, statements, blocks and so on. In the same way, visual languages use compositions to build up the syntactic structure of a picture. Three constituents of a two-dimensional composition are essential to the specification of the visual syntax of a diagram [40]:

1. the *number* of picture elements involved. Binary relationships are the most common, corresponding to the combination of two picture elements. Compositions involving more than two objects are also possible. For example, a connection relationship usually relates to more than two objects. Compositions involving more than two elements can typically be broken down into a series of binary compositions.

2. the *classes* of picture elements involved. The picture element class can be used to restrict objects which are considered for compositions. For example, the containment operator requires a closed picture element rather than an opened element.

3. the *relationship* between the involved picture elements, which can be classified into two different classes. **Coincidence relationships** relate two or more objects based on a common location such as *connect* and *touch*. **Spatial relationships** relate to picture elements by their size and positions such as *contain, align, over, under, left_of, right_of,* and *parallel*.

This work concentrates on compositions which express *connectivity* and *hierarchy* within diagrams. Therefore, the most important composition rules are connection and containment denoted as

$$\Delta = \{\underline{connect}, \underline{contain}\}$$

Connection   *Connection* is a binary composition operator which uses the coincidence relationships between two opened geometrical objects, or an opened and a closed geometrical object. The connection operator considers endpoints of the opened objects and boundaries of the closed objects. For example, a typical connection operator combines two lines where the second endpoint of one line coincides with the first endpoint of the other line. An opened object can be connected with a closed object by defining composition operators for combining opened and closed diagram elements. For example, an endpoint of an opened object falls anywhere

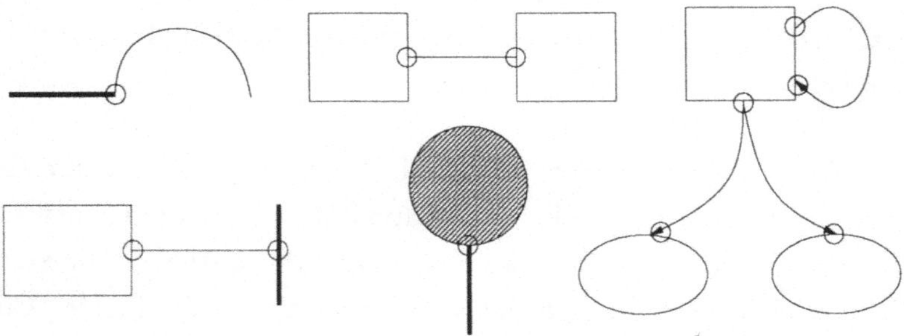

Figure 4.4: Examples of connections

along the boundary of a closed shape. Many additional coincidence relationships
are possible, the relationship can be constrained such as touching in a specific place,
unconstrained such as touching anywhere, or something in between, for example,
touching anywhere along the left boundary. Figure 4.4 illustrates situations where
connections are obviously, as indicated with small circles.

Containment   The boundary of each closed geometrical object defines an area.
If an object is located completely in this area, there is a containment relationship
between these two objects. In other words, containment says that one closed geometrical object encircles another geometrical object. Figure 4.5 gives some examples
for containments which are used for expressing hierarchy.   The large objects which

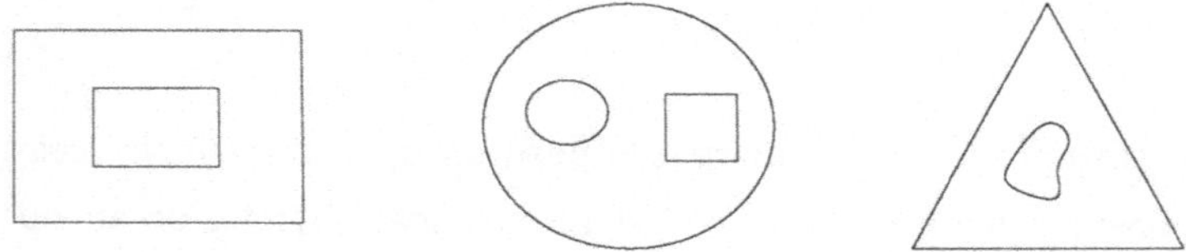

Figure 4.5: Examples of containments

define closing boundaries represent parents, and enclosed objects represent children.
Further, contained objects can contain other objects again.

Alignment   Alignment is used in diagrams to define the relative position between text fragments and picture elements. However, alignment is not used for building the global diagram structure. In a diagram, text information is always attached to a graphical element. For example, the name of a rectangular diagram component can be displayed at the upper left corner of the rectangle symbol, the name of a connection line can be placed somewhere in the middle of the line as shown in figure 4.6.

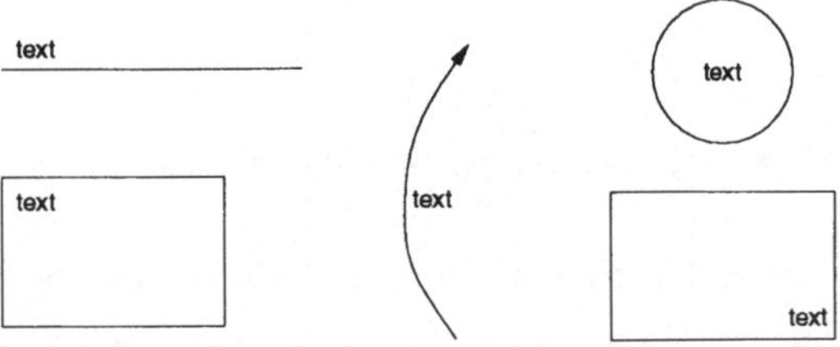

Figure 4.6: Examples of alignments

Pictorial Hierarchy

In definition 4.3, a picture is defined as an unstructured set of picture elements. The following definition introduces a hierarchical structure within such picture elements.

Definition 4.4

A *hierarchical picture* is a set of hierarchical picture elements. A *hierarchical picture element* is a triple, $\hat{e} = (c, A, \mathrm{HP})$, where c is the symbol class, A is a finite set of characteristic attributes, and HP is a set of hierarchical picture elements which are contained in $\hat{e}$ or an empty set.

□

Example 4.1

Figure 4.7 shows a graphical representation of a structured Petri net of which the picture elements are an unstructured picture according to

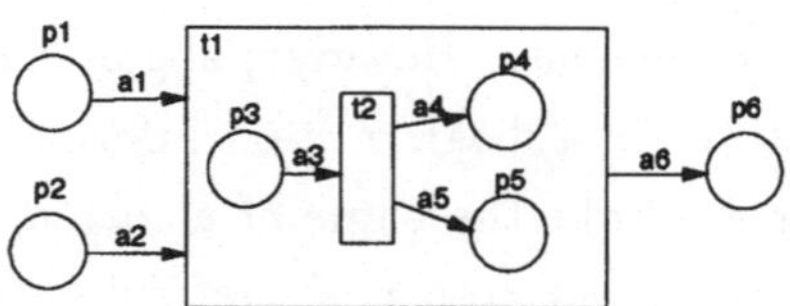

Figure 4.7: A picture of a structured Petri net

definition 4.3:

$$P = \{p_1,\ p_2,\ p_3,\ p_4,\ p_5,\ p_6,\ a_1,\ a_2,\ a_3,\ a_4,\ a_5,\ a_6,\ t_1,\ t_2\}$$

Using the definition 4.4, this unstructured picture can be structured in a hierarchical picture HP by using the containment relationship, so that:

$$\mathrm{HP} = \{\hat{p}_1,\ \hat{p}_2,\ \hat{p}_6,\ \hat{a}_1,\ \hat{a}_2,\ \hat{a}_6,\ \hat{t}_1\} \quad \text{where}$$

$$\hat{p}_1 = (circle, A_{p_1}, \emptyset), \qquad \hat{p}_2 = (circle, A_{p_2}, \emptyset), \qquad \hat{p}_6 = (circle, A_{p_6}, \emptyset)$$
$$\hat{a}_1 = (arrowline, A_{a_1}, \emptyset), \quad \hat{a}_2 = (arrowline, A_{a_2}, \emptyset), \quad \hat{a}_6 = (arrowline, A_{a_6}, \emptyset)$$
$$\hat{t}_1 = (rectangle, A_{t_1}, \{\hat{p}_3,\ \hat{p}_4,\ \hat{p}_5, \hat{a}_3,\ \hat{a}_4,\ \hat{a}_5, \hat{t}_2\})$$
$$\hat{p}_3 = (circle, A_{p_3}, \emptyset), \qquad \hat{p}_4 = (circle, A_{p_4}, \emptyset), \qquad \hat{p}_5 = (circle, A_{p_5}, \emptyset)$$
$$\hat{a}_3 = (arrowline, A_{a_3}, \emptyset) \quad \hat{a}_4 = (arrowline, A_{a_4}, \emptyset) \quad \hat{a}_5 = (arrowline, A_{a_5}, \emptyset)$$
$$\hat{t}_2 = (rectangle, A_{t_2}, \emptyset)$$

The attributes of each picture element describe the characteristic parameters of the corresponding picture element.

For example: $A_{p_1} = \{[0, 4, 1],\ \text{``}p1\text{''}, ...\}$, describes that the picture element $\hat{p}_1$ has the coordinate $(0, 4)$, radius 1, and the text label "p1".

$\triangle$

4.1.1.2 Object Graph

For a textual language, abstract syntax together with the attributing rules defines a class of attributed trees which are used as the internal representations within textual structure editors dedicated to the language. For a diagram language, the directed graph is a natural and theoretically well-founded concept for representing discrete structures [119]. Within a diagram editor, the graph is appropriate for the internal representation.

Basically, a graph contains a set of nodes and a set of edges linking the nodes. Because a diagram usually contains different types of picture elements for representing different types of nodes and edges, we refine this basic concept by classifying nodes and edges into appropriate types. Graphical and other semantic information are represented as attributes attached to nodes and edges.

Definition 4.5

A typed and attributed digraph G is a tuple $(V, E, T_V, T_E, t_v, t_e)$, where

1. V is a finite set of nodes,

2. $E \subseteq V \times V$ is a set of ordered pairs of nodes called *edges*, an edge $e \in E$ which connects the nodes $v_1, v_2 \in V$ can be denoted as $[v_1, v_2]$,

3. T_V is a set of types for nodes,

4. T_E is a set of types for edges,

5. $t_v : V \rightarrow T_V$ assigns a node type to every node,

6. $t_e : E \rightarrow T_E$ assigns an edge type to every edge.

$\square$

A main feature of HiNet diagrams is the representation of the hierarchical structure. Similar to the pictorial hierarchy of external representation, we introduce the definition of *higraph* for the internal representation. The key issue is to introduce so-called hierarchy-nodes to represent hierarchical structures with corresponding hierarchy-edges explicitly.

Definition 4.6

Let $G = (V, E, T_V, T_E, t_v, t_e)$ be a typed and attributed digraph.
$(V, E, T_V, T_E, t_v, t_e, \sigma)$ is a *higraph* iff:

1. $\exists \sigma : V \to 2^V$, σ assigns to each node $n \in V$ a set $\sigma(n)$ of subnodes, and is restricted to be cycle-free,

2. $V_h \subset V$ is a set of hierarchy-nodes with $\forall h \in V_h : \sigma(h) = \emptyset$,

3. $E_h \subset E$ is a set of hierarchy-edges,

4. $\forall v_i \in \sigma(v) : \exists u \in V_h$ with $[v, u] \in E_h \wedge [u, v_i] \in E_h$.

$\square$

Example 4.2

We consider the internal representation of the structured Petri net of which the external representation is shown by figure 4.7. The corresponding higraph is $G = (V, E, T_V, T_E, t_v, t_e, \sigma)$ where:

$$
\begin{aligned}
V = \ & \{p_1, p_2, p_3, p_4, p_5, p_6, a_1, a_2, a_3, a_4, a_5, a_6, top, t_1, t_h, t_2\}, \\
E = \ & \{[top, p_1], [top, p_2], [top, p_6], [top, a_1], [top, a_2], [top, a_6], [top, t_1], \\
      & \ [p_1, a_1], [p_2, a_2], [a_6, p_6], [a_1, t_1], [a_2, t_1], [t_1, a_6], \\
      & \ [t_1, t_h], [t_h, a_3], [t_h, a_4], [t_h, a_5], [t_h, p_3], [t_h, p_4], [t_h, p_5], [t_h, t_2], \\
      & \ [p_3, a_3], [a_4, p_4], [a_5, p_5], [a_3, t_2], [t_2, a_4], [t_2, a_5]\}, \\
T_V = \ & \{Hierarchy Node, Transition, Place, Arc\}, \\
T_E = \ & \{P2A, A2P, T2A, A2P, THierarchy\}, \\
& t_v(a_1) = Arc, \quad t_v(t_1) = Transition, \quad t_v(p_1) = Place, \\
& t_v(top) = Hierarchy Node, \quad t_v(t_h) = Hierarchy Node, \ldots \\
& t_e([p_1, a_1]) = P2A, \ t_e([a_6, p_6]) = A2P, \ t_e([t_1, a_6]) = T2A, \\
& t_e([a_1, t_1]) = A2T, \ t_e([top, p_1]) = THierarchy, \ t_e([t_1, t_h]) = THierarchy, \\
& t_e([t_h, a_3]) = THierarchy, \ t_e([t_h, t_2]) = THierarchy, \ldots
\end{aligned}
$$

Considering σ in definition 4.6, we have:
$$\forall n \in V \setminus \{t_1\} \ : \ \sigma(n) = \emptyset, \quad \sigma(t_1) = \{p_3, p_4, p_5, a_3, a_4, a_5, t_2\}$$

$\triangle$

Higraph Visualization   Figure 4.8 shows a visualization of the higraph G of the above example. This picture illustrates the internal representation by depicting the

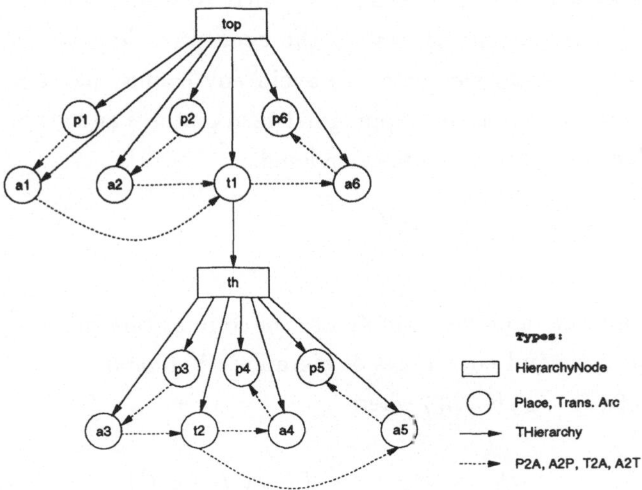

Figure 4.8: Internal representation of a structured Petri net

hierarchy-nodes with rectangles, other nodes with circles; hierarchy-edges with solid arrowlines, and other edges with dashed arrow-lines. It should be noted that the connection lines or arcs of a Petri net are represented internally by node as well. This is similar to the graph grammar approach such as used in [37]. In section 4.4, we use this *graphical notation* for the internal representation of a diagram to illustrate the effects of command interpretation by indicating the corresponding graph manipulations.

4.1.2 Handsketch-based Editing

4.1.2.1 Diagram Schema

In the previous section, we firstly analyzed the picture elements and how these picture elements can be composed to form a diagram. Then we defined the typed and attributed digraph and higraph for the internal representation. This section introduces the definition of a *diagram schema* which brings either representation forms together by the so-called mapping rules. The key issue is to illustrate that there are well-defined *type relationships* between internal graph and external picture. Later on, it will be clear that this *type relationship*, that is, which symbol class represents which type of graph element, is exactly the concept of how to use handsketched symbols to create diagram elements. To avoid any confusions, it is worth to note that we do not try any semantic mapping between pictures and graphs, instead only the visual syntactical structures are considered.

Definition 4.7

Let Π be a visual alphabet and Δ a set of composition rules. T_V and T_E are sets of node and edge types of higraphs. A *diagram schema* DS is a tuple $(\Pi, \Delta, T_V, T_E, R_v, R_e)$ where

$$R_v = \{t \leftrightarrow a \mid t \in T_V \wedge a \in \Pi\}$$

is a set of mapping rules for node representations,

$$R_e = \{t \leftrightarrow c \mid t \in T_E \wedge c \subset \Pi \times \Delta\}$$

is a set of mapping rules for edge representations.

□

A diagram schema brings pictures and object graphs together by saying which symbol class from the visual alphabet represents which type of graph elements. The mapping between node-types and symbol class is simple and unique. However, the mapping rules for edge-representation are more complex than that for nodes because each edge-type concerns both visual alphabet and composition rules. Further, HiNet diagrams have additional restrictions of the mapping rules to enforce a common graphical representation style as follows:

- The picture elements for representing nodes are "closed" graphical symbols.

- Hierarchy is represented by using the visual containment relationship between corresponding picture elements.

- Connection is represented by "opened" symbol such as connection lines which connect the related nodes.

Example 4.3

The example 4.1 and the example 4.2 illustrated a hierarchical picture of a structured Petri net, and an corresponding higraph, respectively. The diagram schema of structured Petri nets is as follows:

$PN = (\Pi, \Delta, T_V, T_E, R_v, R_e)$ where:

$\Pi = \{rectangle, circle, arrowline\}$, $\Delta = \{\underline{connect}, \underline{contain}\}$,

$T_V = \{HierarchyNode, Transition, Place\}$,

$T_E = \{P2TArc, T2PArc, THierarchy\}$

$R_v = \{Transition \leftrightarrow rectangle, Place \leftrightarrow circle\}$,

$R_e = \{P2TArc \leftrightarrow \{arrowline, \underline{connect}\}, T2PArc \leftrightarrow \{arrowline, \underline{connect}\},$
$\qquad THierarchy \leftrightarrow \{rectangle, \underline{contain}\}\}$

$\triangle$

4.1.2.2 Gesture Editing Operator

A diagram schema describes the *type information* of a diagram by specifying the visual alphabet, the compositions, the types of nodes and edges, and the mapping rules between them. In this section, we consider concrete diagrams by the definition of a *diagram configuration* and handsketch-based editing by the *gesture editing operator*.

<u>Definition 4.8</u>

A *diagram configuration DC* is a pair (P, G) which defines a specific diagram by a dual representation of the external picture P and the internal graph G.

An *editing operator op* is a function which transforms one diagram configuration to another diagram configuration, denoted by $DC_2 = op(DC_1)$. Each *op* can be denoted as $op_g \parallel op_p$, where op_g manipulates the internal representation, and op_p manipulates the external picture.

$\square$

A diagram editor consists of a set of editing operations for creating and modifying diagrams. The key point here is the duality, that is, an editing operator operates simultaneously on the external picture and the internal graph which are mutually dependent.

Definition 4.9

An *editing process* is represented by a sequence of diagram configurations.

$$DC_0, DC_1, DC_2, ..., DC_n \text{ where } DC_{i+1} = op_i(DC_i)$$

DC_0 is the start diagram configuration and DC_n is the diagram configuration after n editing operations.

□

Similar to Arefi's approach [5] which provides a mechanism to unify the specification of the language and its manipulations, we consider a visual language as an initial object and a collection of editing operations. Any object that can be obtained by applying a sequence of allowed editing operations is then defined to be in the language. Therefore, a visual program can be constructed from one diagram configuration to another by applying appropriate editing operations as shown by definition 4.8. The syntactical correctness of the edited diagram is guaranteed by the correctness of each editing operation. This was also considered by Goettler [38], who indicated that the question of syntactically correct editing can be reduced to "What is allowed to be in the diagrams?"

Definition 4.10

A gesture operator is a triple $\eta = (s, C, op)$ where s is the symbol class of a handsketched picture element, C is a set of constraints which the picture element must fulfill, and *op* is an editing operator as defined before.

□

This definition is the formal basis for the concept of our gesture specification mechanism by specifying a gesture command in its shape, its constraints, and its semantics. As a matter of fact, each η corresponds to a gesture class. Finally, the following definition builds the formal basis of our concepts for the high-level recognition which will be discussed in the next section in full detail.

Definition 4.11

A *handsketch-based diagram editing system* S is a tuple $(DS, \mathcal{G})$ where

1. $DS = (\Pi, \Delta, T_V, T_E, R_v, R_e)$ is a diagram schema and $\mathcal{G}$ is a set of gesture operators.

2. $\Pi \subset G_s$, $\quad G_s = \{ s \mid (s, C, op) \in \mathcal{G} \}$ is the set of symbol classes used in gesture operators.

$\square$

In section 4.4, we describe the classification of editing commands. At this point, it is only important to note that the symbol classes in Π represent only the constructive commands for creating a diagram. There are other gesture operators for manipulating diagrams, and their gesture shapes are usually not in the visual alphabet Π.

A diagram schema DS defines the *type information*, and, the constraints of each gesture operator in $\mathcal{G}$ guarantee the syntactical correctness of each transformation from one diagram configuration into another. Indeed, there is a *generation relationships* between a diagram schema DS and a specific diagram configuration DC, which is not formally defined in this work due to the scope of required formalism.

The key issue of this definition is that the visual alphabet of the diagram schema corresponds to the set of symbol classes used in the gesture operators. Further, if a gesture operator (s, C, op) creates a diagram component, the picture element is exactly from the symbol class s. The following example illustrates this definition.

Example 4.4

PNS $= (PN, \mathcal{G})$ is a Petri net editing system, where PN is Petri net schema as described in example 4.3. $\mathcal{G} = \{\eta_1, \eta_2, \eta_3, \eta_4\}$ is the set of gestures for editing Petri nets, where

$\eta_1 = (circle, \{\text{nonoverlaps}\}, op_{createPlace})$

$\eta_2 = (rectangle, \{\text{nonoverlaps}\}, op_{createTransition})$

$$\eta_3 = (opened, \{\text{connect a circle and a rectangle}\}, op_{createArc}\}$$
$$\eta_4 = (cross, \emptyset, op_{delete}\}$$

The Petri net editing system PNS defines four gesture operators for creating and manipulating Petri nets. The effects of each gesture operator will be discussed in section 4.4.

$\triangle$

4.2   Fundamental Concepts

4.2.1   Compound Specification

A handsketch-based diagram editor is a graphical structure editor. The editing objects are not just graphical symbols such as in the case of a general drawing editor, they are rather well-defined diagram elements. Therefore, the underlying diagram language must be defined.

A significant feature of handsketch-based diagram editors is that gestures embody several *compound* information such as the user interface in form of handsketches, the underlying diagram syntax in form of constraints, and editing operations together with all of the required parameters. This was indicated by the definition of the gesture operator. There are well-established methods for specifying textual languages, but this is not the case for visual languages. Our basic idea to specify the language-dependent high-level recognizer is to define a set of gesture operators, each of them consists of a gesture shape, a set of gesture constraints, and gesture semantics. In the following, we use the handsketch-based Petri net editor, one of our experimental applications, as an example to illustrate the most important design issues and considerations as well as some suggestions.

4.2.1.1   Gesture Shapes

A gesture shape is a symbol class whose picture element can be handsketched and recognized by the low-level recognizer. Each gesture class defines a gesture shape as the reference symbol for matching. As aforementioned in section 3.4.5, there are two basic types of graphical symbols, simple symbols and composite symbols. Symbols which are defined directly in the symbol hierarchy are simple symbols. Symbols which consist of several simple symbols are composite symbols.

Our handsketch-based Petri net editor has ten gestures for creating and manipulating Petri nets. Table 4.1 shows the set of gesture shapes we chosen. The gesture names are self explanatory, and all gesture shapes are symbol classes which are directly supported by the low-level recognizer. In section 7.2, the implementation details of how to define gesture shape are described.

Table 4.1: Gesture shapes defined for editing Petri nets

Gesture name	Gesture shape
CreatePlaceGesture	*Ellipse*
CreateTransGesture	*Quadrilateral*
CreateArcGesture	*Opened*
SelectGesture	*Ellipse*
DeleteGesture	*Cross*
ActivateTransGesture	*SharpArrow*
MoveGesture	*SharpArrow*
AddTokenGesture	*Dot*
NameGesture	*HorizontalLine*
ClearGesture	*ZForm*

The main issues for designing gesture shapes are as follows:

- The shape should be intuitive and easy to learn. For "create" commands, the graphical symbol of the object to be created should be used as the gesture shape in the light of our input principle of "what you draw is what you get" (WYDIWYG). The cross symbol is the standard gesture shape of the delete command, which is abstracted from the paper and pen metaphors of everyday life.

- The shape should be chosen as general as possible to improve the recognition rate by tolerating the user's drawing mistakes. For example, using an ellipse as the gesture shape for "create place" improves the recognition rate, because the user usually means a circle but draws an ellipse. For the same reason, the shape for "create transition" gesture uses quadrilaterals instead of rectangles. However, the recognizability of each shape must be considered. In case that both parallelograms and rectangles are used in a diagram language with the same constraints for different object types, quadrilaterals are ambiguous, and can therefore not be used.

- The shape should contain enough geometrical information which the corresponding gesture command needs. For example, the "add token" gesture needs only a point to indicate where the token should be inserted, so that a dot is enough. In contrast, the "create transition" gesture requires not only the position but also the size of the transition, so a rectangle can be used as the gesture shape.

4.2.1.2 Gesture Constraints

Matching handsketches with each gesture shape builds the front-end of the high-level recognition. However, the meaning of a gesture shape is similar to type information in a textual programming language, which is not enough to guarantee the syntactical correctness of a gesture command. Further, there are usually several gesture shapes which belong to the same symbol class which cannot be discriminated one from another without additional information. In our Petri net example, "SharpArrow" is used both as the gesture shapes for "activate transition" gesture and for "move objects" gesture. The high-level recognizer can only recognize a "SharpArrow" by considering additional *conditions*. It is a "activate transition" gesture in case that the arrow is drawn over an enabled transition object; and it is a "move objects" gesture, if the gesture is not drawn over an enabled transition, and there are selected objects. Such conditions which make a gesture *valid* are called *gesture constraints*. Besides such "discrimination conditions", gesture constraints mainly deal with structure recognitions which will be discussed more detailed in section 4.3.

Table 4.2 gives a short overview of the constraints defined for Petri net editing gestures. *None* means that no gesture constraints are defined for that gesture. Gesture constraints are expressed directly in the implementation language C++. It is worth noting that the constraints for CreatePlaceGesture and for CreateTransGesture depend on the type of the Petri net to be edited. In section 4.3, we discuss the additional constraints for recognizing the hierarchy of transitions in hierarchical Petri nets. Our strategy for simplifying the definition and realization

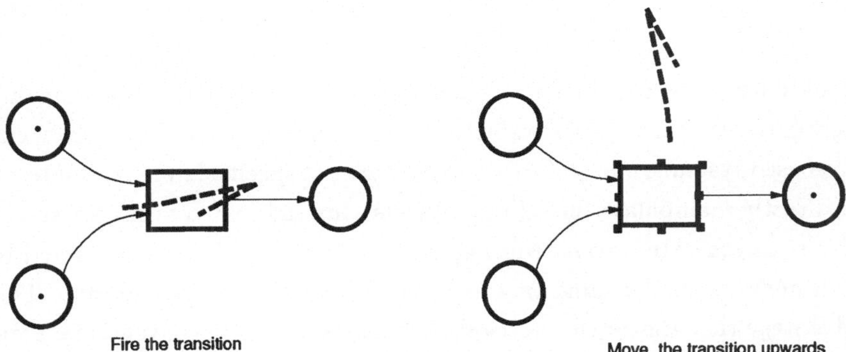

Figure 4.9: Gesture constraints make possible to use the same gesture shape for different gesture commands.

of gesture constraints is to provide a set of basic *constraints operations* which can be directly reused to define new constraints. These operations deal with the most frequently used examinations for spatial relationships, they build the basic protocols of the Handi architecture which are discussed in chapter 5.

Table 4.2: Gesture constraints defined for editing Petri nets

Gesture name	Constraints
CreatePlaceGesture	Gesture does not intersect any places or transitions.
CreateTransGesture	Gesture does not intersect any places or transitions.
CreateArcGesture	One endpoint of the gesture connects a place and the other endpoint connects a transition.
SelectGesture	None
DeleteGesture	None
ActivateTransGesture	Gesture intersects an enabled transition.
MoveGesture	Gesture does not intersect an enabled transition, and there are objects selected.
AddTokenGesture	Gesture is inside a place.
NameGesture	Gesture intersects a name label.
ClearGesture	None

4.2.1.3 Gesture Semantics

One goal in designing our high-level recognizer is to permit easy integration of a gesture recognizer within an object-oriented editor architecture. Different from existing gesture-based systems, our gesture semantics are not specified in form of interpreters which directly manipulate the editing objects. Instead, our concept for gesture semantics is to generate *normal* editing commands which will be interpreted by the editor framework in the same way as in conventional graphical editors. This has the advantage that the gesture recognizer is *compatible* with conventional graphical structure editors. Further, the same editing commands can be invoked both by using gestures and by using menu buttons.

An editing command usually needs parameters, such as a move command requires the destination position to which the selected object should be moved. In conventional graphical user interfaces, command parameters are collected by using direct manipulation techniques such as dragging or by using dialog forms to prompt user's input. In contrast to this kind of *explicit* parameter collection, gesture commands have the property that the required command parameters such as "position" or "which object" are contained implicitly in the handsketch or in the spatial relationships between the handsketch and the external representation of the underlying diagram. For example, if one uses an arrow symbol as the move gesture, the destination position can be calculated from the size and position of the handsketched arrow. Therefore, the main task of the command generation is to recognize the command parameters and to create appropriate editing commands.

Table 4.3 informally describes the gesture semantics by saying what each command does. The "gesture contained" command parameters are emphasized with "**the**" in the command descriptions.

It is important in the design of each gesture to consider that some required command parameters can be recognized from the handsketches during the constraints checking process. For example, in checking gesture constraints defined for `CreateArcGesture`, the place and the transition which should be connected can be found and stored while the constraints are examined. In this way, these information can directly be used as command parameters in dealing with the gesture semantics. The recognition process is therefore more efficient because the information which is used by the gesture semantics is already prepared in checking the gesture constraints.

Table 4.3: Short description of gesture semantics used for editing Petri nets

Gesture name	Short descriptions
CreatePlaceGesture	Create a place at **the** position
CreateTransGesture	Create a transition at **the** position
CreateArcGesture	Create an arc to connect **the** place and **the** transition
SelectGesture	select objects
DeleteGesture	select objects and remove selected objects
ActivateTransGesture	select and fire **the** enabled transition object
MoveGesture	Move selected objects
AddTokenGesture	Add a new token into **the** place
NameGesture	Change **the** text
ClearGesture	Clear all strokes

Since a single gesture can contain several editing commands with all required parameters, the gesture semantics are able to generate so-called macro commands which consist of several single commands.

4.2.2   Object-Oriented System Design

Figure 4.10 shows the components and their organizations of the high-level recognition system. The input of the high-level recognizer is formed by graphical symbols recognized by the low-level recognizer, and the output of the high-level recognizer is a stream of editing commands. Within an object-oriented design, both graphical symbols and editing commands are objects. Modeling editing commands as objects is similar to other operation-as-objects approaches characterized by Meyer [80].

The core of the high-level recognizer is a set of gestures which are defined as described in the previous section. Whenever the user has drawn a new symbol, each gesture uses the selective matcher supported by the low-level recognizer to match the gesture shape with the current handsketches. If the symbol matches the handsketches and all constraints are fulfilled, the appropriate editing commands are created. The parameters which the commands require are usually collected by the constraints checking process.

Within our object-oriented design, gestures are modeled as objects instanced from appropriate gesture classes according to our specification mechanism. The

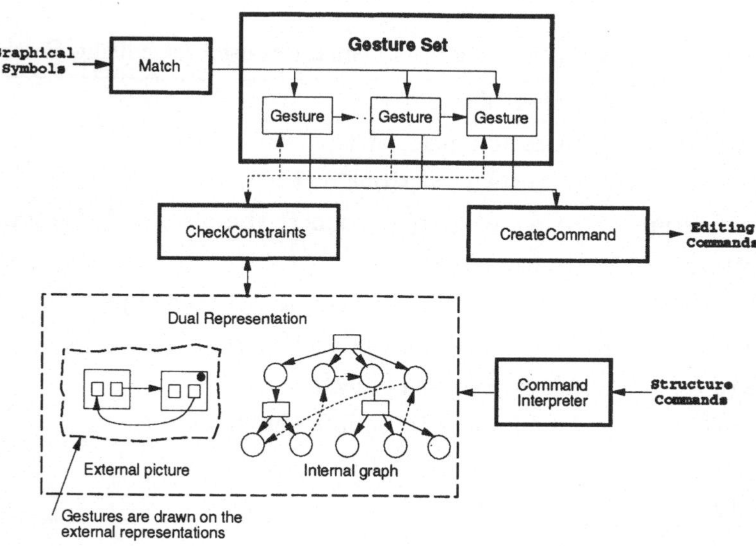

Figure 4.10: The high-level recognition system

matching of gesture shapes, the checking of gesture constraints, and the creating of editing commands are designed as *virtual functions* [11] which can be invoked by the same interface. This polymorphism technique of the object-oriented design supports a unique access to methods of matching, of constraints checking, and of command creation. This is illustrated in figure 4.10 by using three blocks which represent these virtual methods. The following pseudocode describes the control structure of the algorithm used by the high-level recognizer.

```
for gesture = gestureSet.first() until gestureSet.last() do
    if gesture.shape is matched by the low-level recognizer
    then if gesture.CheckConstraints() returns OK
        then return gesture.CreateCommand()
        end if
    end if
    gesture = gestureSet.next()
end for
```

As defined in definition 4.8, a diagram is represented simultaneously in an internal graph and an external picture. A gesture is always drawn in relation with the external picture, that means, the spatial relationships between a handsketched gesture and the external graphical representation are used to define the diagram syntax. This is expressed in form of constraints, and realized in form of methods for examining spatial relationships between the external representation of the underlying diagram and the considered handsketch.

Conceptually speaking, the functionality of a high-level recognizer is limited to the transformation of handsketches into editing commands. The execution or the interpretation of an editing command belongs to tasks of the underlying editor framework. However, the interpretation of *structure commands* requires syntactical knowledge of the underlying diagram. Within an object-oriented editor, editing commands are objects which are delegated to objects which are responsible for interpreting diagram-specific structure commands. Therefore, as indicated in figure 4.10, *structure commands* which are originally generated by the high-level recognizer come back to the diagram components for interpretation. Other editing commands such as zooming are interpreted by the used editor framework which are not considered in this work.

The functionality of the command interpreter contains parsing of new created diagram elements and required structure manipulations. Each editing command simultaneously manipulates on the internal representation and the external graphical representation of the underlying diagram.

4.3   Structure Recognition

This section deals with the key problem of the high-level recognition, that is, the structure recognition. A HiNet diagram consists of mainly two kinds of structures, namely the *connectivity* and the *hierarchy*. The guiding principles for recognizing these diagram structures are the concepts of the *diagram schema* and the *mechanism* for the gesture specification which are presented in the previous sections. The symbol class of a recognized handsketch identifies the type of the corresponding graph element. Within the gesture constraints, the diagram syntax can be examined in full detail. Therefore, the key issue of the structure recognition is done by considering the appropriate gesture constraints.

The main task for checking gesture constraints is to examine the spatial relationships between the handsketch which matches the considered gesture shape and the external representation of the underlying diagram. In the following sections, we discuss issues for recognizing the hierarchical and the connective structures by using a number of examples.

4.3.1 Hierarchy

Before beginning the discussion, the term of *hierarchy* needs some explanation to help avoiding confusions. The term hierarchy is used in this work for *visual* and *syntactical* structure. It is important not to confuse hierarchies such as refinement hierarchy in system design with nets [30] or class hierarchy in object-oriented programming [11]. Usually, these hierarchies are represented visually in trees instead of using the containment relationship. On the contrary, for example, Venn diagrams use our syntactical hierarchy structure to express the subset semantics. This work considers hierarchy as a syntactical structure which is visually represented by using the containment relationship.

The recognition of the hierarchical structure is to examine the containment relationship between the actual handsketch and the graphical representation of the already recognized diagram fragments. Simply stated, it is to find the smallest node object which contains the sketched object. The external representation of a diagram is organized in a hierarchical manner according to the pictorial hierarchy introduced by the definition 4.4. This provides an efficient structure for searching desired graphical objects because our pictorial hierarchy corresponds to the containment hierarchy of the graphical objects.

Examples

Figure 4.11 illustrates a handsketch-based editing scenario where the user drew a rectangle which is a "create transition" gesture. Within the high-level recognizer, the gesture shape of the "create transition" gesture matches this handsketched rectangle, the responsible gesture constraints are examined. This handsketched rectangle does not intersect with any other diagram objects, and it is found that the rectangle of the transition t2 is the smallest transition which contains this handsketch. Therefore, the transition t2 is the hierarchy-parent.

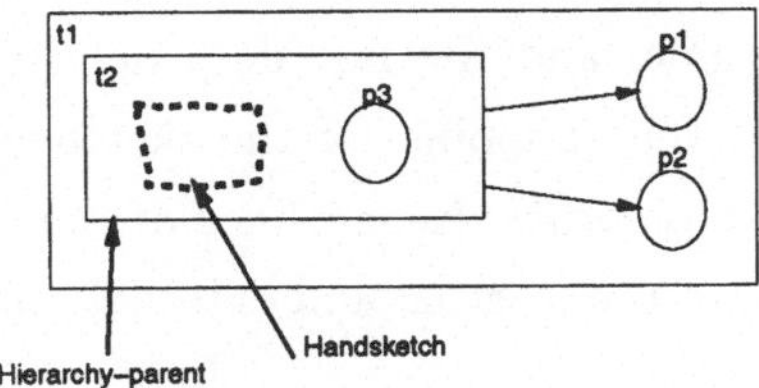

Figure 4.11: Recognizing the hierarchical structure with a structured Petri net

Figure 4.12 illustrates a more complex example under the same context in recognizing hierarchical structures. In this example, the smallest rectangle which contains the handsketched rectangle is that of the transition t1. Therefore, the transition t1 is the hierarchy-parent. Further, the handsketched rectangle contains existing Petri net components, the existing hierarchical structure must be changed. It is obviously that the transition t2 will be a subnode of the new transition. The manipulation of the diagram structure is the task of the command interpretation which will be discussed in the next section in full detail. However, these information investigated in checking gesture constraints are stored in the corresponding gesture object, and used in interpreting gesture semantics.

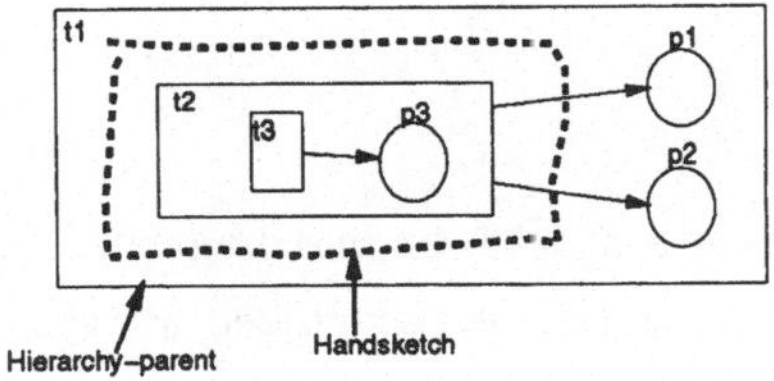

Figure 4.12: Inserting a new node can change the existing hierarchy structure.

Additionally to that "node-sketches" can invoke hierarchy manipulations, hierarchy can also be created by diagram-specific gesture commands. Figure 4.13 illustrates this by two statechart examples. Statecharts provide an interesting visual representation of orthogonal states by using dashed lines. The orthogonal states and the parent xor state build hierarchical structures between states. In our example, handsketched lines are gestures for creating orthogonal and-states. The gesture constraints are defined as follows: The sketched line is drawn inside a state, it is a horizontal or a vertical line, and the line does not intersect with other object inside this state. Further, the endpoints of the sketched line must coincide with the boundaries of the state or other dashed lines of the considered state. The required manipulations both of the internal and external representation are described in section 4.4

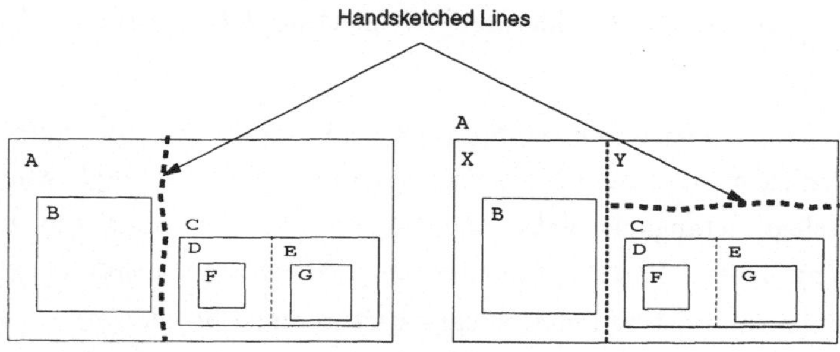

Figure 4.13: Recognizing hierarchy of orthogonal states in a statechart

4.3.2 Connectivity

A connective structure depends at least on one diagram element which belongs to the node type. Therefore, a connection is recognized always in two steps: recognizing the nodes to be connected and recognizing the connection lines. The recognition of nodes is done by matching the gesture shapes and subsequently by checking additional constraints as discussed before.

In contrast to the hierarchical structures based on the containment relationships, a connective structure is based on the coincident relationships. Therefore, the gesture constraints defined for a connection gesture deal always with the two endpoints of a handsketched line. The concrete constraints express diagram syntax in form of where the endpoint of a connection line can be positioned. Further, the type of the related diagram elements can be used to constrain the type of the connection. This is particularly useful in case that the same gesture shape is used for different connection types.

The most important task to recognize a connection is to find node objects at the endpoints of the handsketched line, and to check the additional semantic constraints such as places can only be connected with transitions in Petri nets. For efficient searching of *connectable* diagram elements, each node has invisible connectors in areas where a connection can exist. To tolerate the inaccuracy of handsketches, two small invisible circles around the endpoints are used in the searching operation as illustrated in figure 4.14.

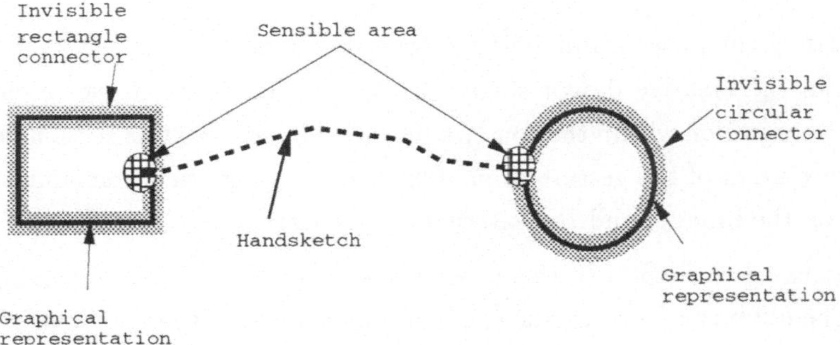

Figure 4.14: Connectors support to recognize the connection structure.

4.4   Command Interpretation

Within a handsketch-based diagram editor, editing commands can be classified into two global classes: structure manipulating commands and structure preserving commands. The interpretation of structure preserving commands such as zooming is supported by the used editor framework, which is not the topic of this dissertation. Structure manipulation can further be classified into *constructive* operations and *destructive* operations. Constructive commands include various "create" commands for different diagram components. Destructive command is mainly the "delete" command which must be interpreted by considering the consistency of the underlying diagram structure.

In section 4.1.2, we have introduced the definition of diagram configuration. An editing process is characterized by a sequence of diagram configurations which are transformed from one into another. A diagram configuration can be visualized by illustrating the external picture and the internal graph. In the following sections, we describe the gesture semantics of each gesture editing command by showing the diagram configuration <u>before</u> the interpretation and the diagram configuration <u>after</u> the interpretation.

4.4.1 Constructions

The most frequently used commands in editing diagrams are constructive commands, specially in the creative design stage. In the previous section, we discussed the structure recognition within the constraints checking process, this section deals with the interpretation of the gesture semantics. Each editing operation simultaneously operates on the internal and the external representation.

As discussed in chapter 2, this process is similar to incremental parsing of diagrams. The difference between the command interpreter of the high-level recognizer and a diagram parser is that the inputs of the command interpreter are not rough pictures elements, rather they are syntactically correct editing commands. This is because that the gesture shape and the gesture constraints are examined before the gesture semantics are considered. The main task of the command interpretation is to execute the required graph manipulations. The interpretation of constructive commands is to add the just created objects correctly into both the internal and the external representation.

An Example

To illustrate how to interpret "create" commands, we consider a concrete example of a hierarchical Petri net. Figure 4.15 shows the interpretation of a sequence of constructive commands one by one. The left part of this figure depicts the actual display of the external graphical representation, and the right part shows the internal graph representation. Each line of the figure represents a diagram configuration, and the handsketches illustrate the gesture commands drawn directly on the external representation. The effect of the command interpretation becomes obviously by comparing the the internal and external representations before and after the manipulations.

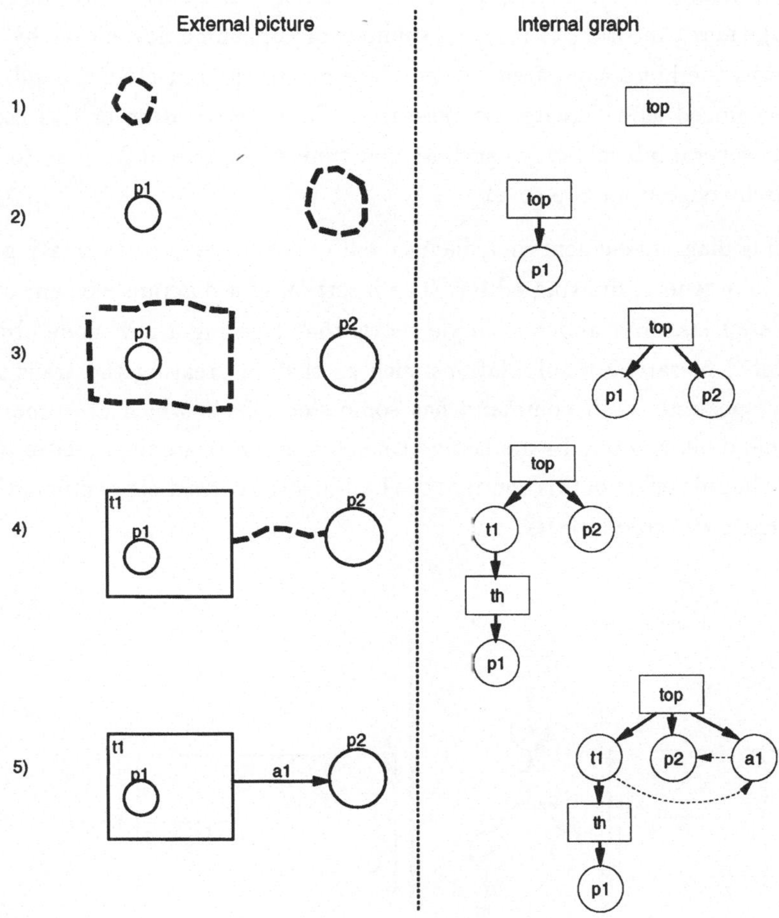

Figure 4.15: Interpretation of constructive commands

At the beginning, the drawing area is empty, and there is an initial hierarchy-node **top** in the internal graph. In the second step, one draws a place, this is recognized as a "create place" command. The interpretation of this "create place" command produces a place object as a child of the hierarchy-node **top**. Similarly, another place object is created under this hierarchy-node. Then the user draws a rectangle which is the gesture to create transition objects. In the process of checking gesture constraints, this handsketched rectangle is recognized as the hierarchy-parent of the place **p1** because this rectangle contains the circle of the place **p1**. In interpreting this gesture command, the third diagram configuration is manipulated into the fourth as follows: the new transition `t1` is now a subnode of `top`, and a new hierarchy node `th` is created as the hierarchy-parent of the place `p1`. In the last step, a handsketched line is recognized as a "create arc" gesture. The interpretation of this command constructs several internal edges such as that from `t1` to `a1` and from `a1` to `p2`, and the hierarchy edge from `top` to `a1`.

Creating diagram elements in a diagram editor is not the same as creating picture elements in a general drawing editor. The insertion of a diagram element can have many *side effects*. The above example shows that creating a new node object can invoke global hierarchy manipulations such as that in creating the transition `t1`. Generally speaking, each command has some side effects which are produced by the command interpreter. In the last editing step of our example, an edge object is created. The side effect of this command is building the correct connection references between the connected objects.

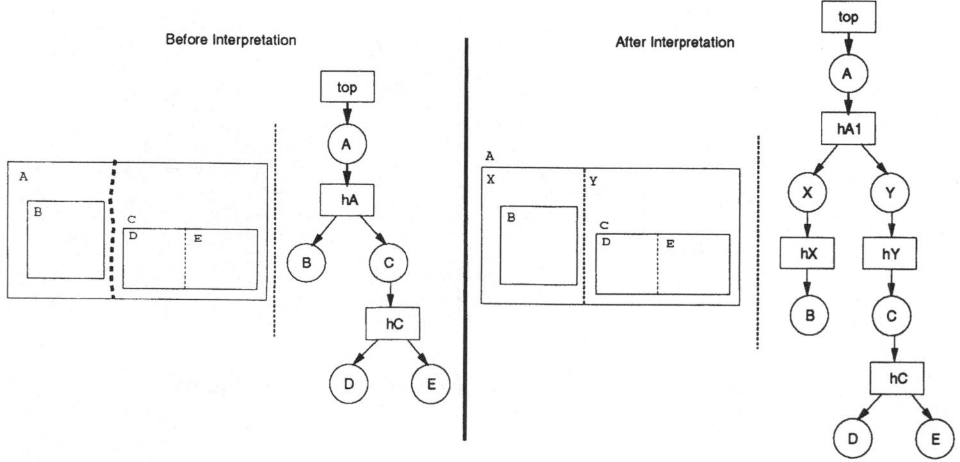

Figure 4.16: Interpretation of a statechart-specific command

Diagram-Specific Operations

Each individual diagram has some specific features, and the corresponding gestures must be interpreted with special considerations. Figure 4.16 describes the interpretation of the command used in a statechart editor as aforementioned in the last section. The handsketched line is recognized as an "xor to orthogonal" command, and the required manipulations both in the internal and in the external representation are illustrated by showing the diagram configurations before and after the interpretation. The left-hand side of the figure depicts the original diagram configuration, and the right-hand side depicts the diagram configuration after the interpretation. Externally, the handsketched line is changed to a dashed line according to the graphical syntax of statecharts in representing orthogonal states. Internally, the original xor-hierarchy-node hA of state A is changed to hA1 of the type of and-hierarchy-node. Two orthogonal states X and Y and two corresponding hierarchy-nodes hX and hY are created. Originally, the state B and C are subnodes of the state A, after the interpretation, they are subnodes of the state X and Y, respectively.

4.4.2 Destructions

One of the most frequently used commands in a conventional graphical editor is the select command. To delete an object, the user first has to select it by using a select command. Further, object-oriented commands can only perform operations on the selected objects. Many structure-oriented navigation commands such as that for wandering through the internal graph support a convenient interface to select desired objects.

In contrast to conventional graphical structure editors, the select command is usually an implicit command in a handsketch-based editor. This is in particular obvious in the "delete" gesture. Within a handsketch-based editor, the user draws a cross symbol over the object which should be deleted, without selecting this object before. This is realized by the concept of so-called macro commands which combines several single commands together. Therefore, the gesture semantics of a "delete" gesture is a macro command which includes a select command and a delete command. The geometrical information of a handsketch such as the position and the size can always be used for selecting objects. The interpretation of a macro command is done by interpreting each single command one after another, such concept is supported by most editor frameworks.

The select command is interpreted by the editor in a conventional way. The interpretation of a delete command differs from deleting picture elements in general drawing editors. Diagram components, which the user wants to delete, may have syntactical relationships to other diagram elements. The interpretation of a delete command includes all updates to guarantee the consistency. For example, if a node should be deleted, it is necessary to delete all edges which are connected with this node in order to maintain the consistency. Similarly, if an edge should be deleted, the references which are managed in connected nodes should be removed as well.

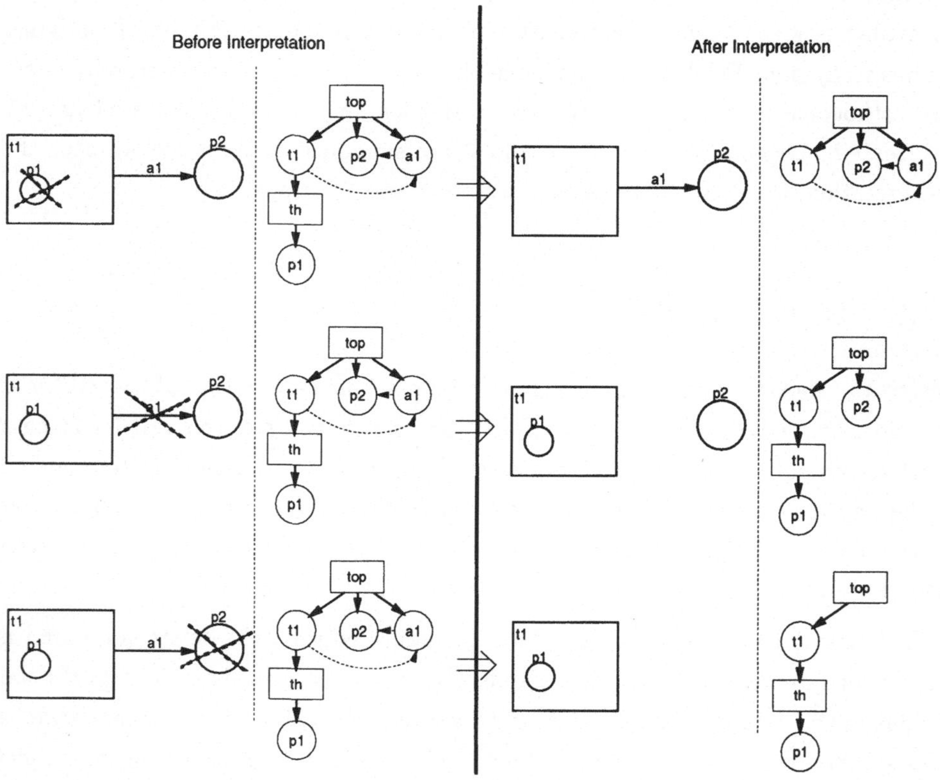

Figure 4.17: Interpretation of delete commands

Examples Figure 4.17 illustrates the interpretation of several delete commands. A handsketched cross symbol depicts the delete gesture. The left-hand side shows the diagram configurations before the command interpretation, and the right-hand side shows the diagram configuration after the interpretation. The effects of each gesture delete command become obviously by comparing the differences between the diagram configurations. Deleting a node with subnodes invokes a hierarchical destruction, that is, to remove all subnodes.

4.5 Summary

After the formal introduction of HiNet diagrams and handsketch-based editing, we presented the mechanism for specifying gestures in gesture shape, gesture constraints, and gesture semantics, as well as our object-oriented design of the high-level recognition system. The gesture shapes are graphical symbols which can be recognized by the low-level recognizer, the gesture constraints deal with the visual compositions, and the gesture semantics are formed by editing commands. Different from the low-level recognizer, the high-level recognizer is designed by using the specific characteristics of HiNet diagrams such as containment representing hierarchy and connection lines representing connectivity. However, our basic concept can be generalized by concerning other spatial relationship such as above, left, right, or touching to recognize structures used in other visual languages.

Chapter 5

Handi Architecture

This chapter presents the Handi architecture which combines the low-level recognition and the high-level recognition as integrated components of handsketch-based diagram editors. Handi stands for "**Han**dsketch-based **di**agram editing." We begin with the design goals and an overview of the architecture, outlining its major elements, and showing the mechanism which assembles the elements. Then we consider the Handi's subsystems in detail, describing their semantics and relationships by using class diagrams and object diagrams. We conclude the chapter with a summary of the architecture.

5.1   Introduction

5.1.1   Motivation and Design Goals

The development of handsketch-based diagram editors is difficult. There are basically two reasons for this:

1. Gesture specification and gesture recognition is hard. The most important components of handsketch-based diagram editors are the recognition components. Recently, handwriting recognition has been integrated in several pen-based operating systems such as PenPoint [17] and Windows for Pen Computing [21]. However, there are only a few gesture-based graphical applications. It lacks of concepts of gesture specification and gesture recognition for handsketch-based diagram editors.

2. Currently available editor frameworks and user interface toolkits are designed for a broad range of domains such as technical and artistic drawing, music composition, circuit design, and many others. Such editor frameworks have to resign many common features of diagram editors because of the endeavor of the generality. Using these general editor frameworks, the editor developer must design these common features for each individual editor repeatedly, which is a very time consuming task.

Once complexity reaches a certain level in a software system, a new level of abstraction is necessary to allow further increases in functionality. The development of a new level of abstraction relies on gaining enough experience with a class of applications so that their implementations are well understood. The design of Handi follows exactly this basic principle. We built several prototype editors for examining the application class of handsketch-based diagram editors to ascertain the common fundamental elements within such editors. The basic abstractions are developed by extracting common elements and encapsulating them into reusable classes.

The primary goal of this research was the design of the Handi software architecture which reduces the development efforts for handsketch-based diagram editors. On the one hand, Handi specializes basic abstractions of general editor frameworks for the purpose of diagram characteristics, and on the other hand, Handi provides new abstractions considering gesture specification and gesture recognition. The recognition algorithms developed in the previous chapters are encapsulated into classes which can be used easily.

The design of Handi focuses on creating a software system with the following key attributes:

1. It supports handsketch-based diagram editing.

2. It significantly reduces the time and effort needed to develop handsketch-based diagram editors.

3. Handi-based editors have unique gestural interfaces which are easy to use and support the visual programming environment.

5.1.2 Overview

As stated in the introduction, this work does not treat visual languages in general, instead, the concentration is situated on diagram languages. In contrast to the repeatedly investigated grammar and generator approach both on structure editors and on spatial parsers, our research strategy is to design an object-oriented framework.

A fundamental design decision in Handi was to adopt an object-oriented model in which classes encapsulate common attributes. The aim is to encapsulate the common characteristics of diagram editors into basic programming abstractions. A new handsketch-based editor can be built on top of these basic abstractions, reusing the common implementation, and therefore, reducing the development time for each individual editor.

The classes which encapsulate the common features of handsketch-based diagram editors form the software layer Handi. Figure 5.1 depicts the dependencies between the layers of software that underlie a handsketch-based diagram editor based on Handi. At the lowest levels are the operating system and the window system. Above the window system level, a general purpose editor framework and a user interface toolkits are located. Handi stands at the highest level of system software, providing abstractions that are closely matched to the requirements of handsketch-based diagram editors.

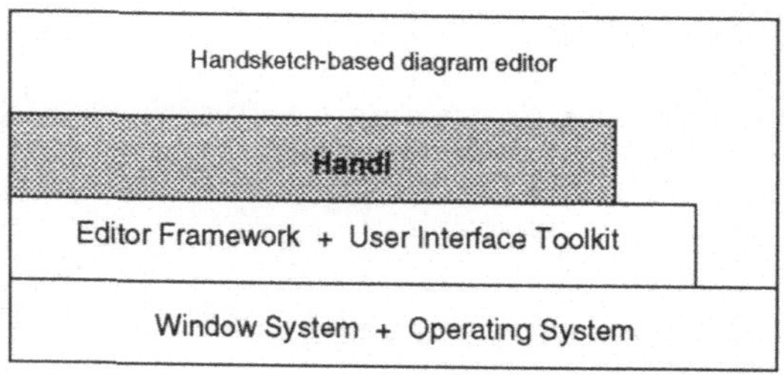

Figure 5.1: Relevant layers of Handi-based editors

An editor for a particular diagram language relies on Handi for its handsketch-based editing capabilities and the basic implementation of common features of diagram editors, on the editor framework and user interface toolkit for their general editing commands and the "look and feel" of the user interface, and on the window and operating systems for managing workstation resources.

Considering the language specification, Handi provides basic abstractions such as internal and external representations of hierarchy, node, and edge objects. The editor developer can specify a new diagram by using these basic abstractions, reusing the fundamental implementation, and adding just the new features. Therefore, Handi reduces the implementation requirements of handsketch-based diagram editors.

5.1.2.1 Subsystems

In designing the Handi architecture, we focused on the common attributes of handsketch-based diagram editors. The software components of Handi consist of the following subsystems as shown by figure 5.2.

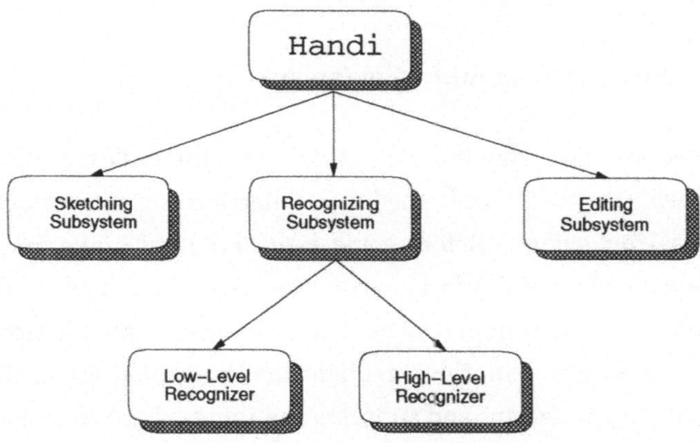

Figure 5.2: Handi consists of three subsystems.

1. The **sketching** subsystem provides an input model for handling gestural inputs. In contrast to the conventional low-level inputs in form of various events, gestural inputs are built in form of handsketches. This subsystem manages such input, performs inking of the stylus trace, and converts the input device manipulations into *stroke objects* which encapsulate attributes of each handsketch.

2. The **recognizing** subsystem consists of the low-level recognizer and the high-level recognizer corresponding to the concept for gesture specification and gesture recognition which have been presented in the previous chapters. Handi provides the basic gesture abstraction which can be subclassed for each individual application.

3. The **editing** subsystem of Handi provides basic abstractions for representing and editing diagrams. Similar to the Smalltalk MVC mechanism, which is used in most object-oriented editor frameworks, we relate the internal graph to the *model* and the external graphics to the *view* of each diagram component. Common characteristics of diagrams are encapsulated into basic classes such as hierarchy, node, and edge classes which can be subclassed for each specific diagram language. While the components interpret editing commands, the corresponding view classes provide utilities for checking spatial relations between gestures and diagram elements, that are used by the recognizing subsystem.

5.1.2.2   Key Abstractions and Mechanisms

Within the object-oriented Handi architecture, the basic building blocks are classes and objects. Each subsystem is designed as a collection of classes which encapsulate proper states and behaviors. Before going into details, we give first an overview about the key abstractions which form part of the vocabulary of our problem domain. As Booch [11] states, "The primary value of identifying such abstractions is that they give boundaries to our problem; they highlight the things that are in the system and therefore relevant to our design, and suppress the things that are outside the system and therefore superfluous." After identifying key abstractions which form a model of reality, mechanisms must be designed by adding behaviors to these abstractions. Whereas key abstractions reflect the vocabulary of the problem domain, mechanisms are the soul of the design that consider how instances of basic abstractions work together.

Figure 5.3 illustrates the key abstractions and mechanisms of Handi by figuring the relationships between key objects. It should be mentioned that the figure is not complete and rather abstract, it depicts only the most important states and behaviors. Some control flows refer to the underlying editor framework, which can only be clarified by the implementation details. One of the most important aspects in object-diagrams is the sequence of message sending or the timing. Similar to the notation used in [129], the numeric labels in the figure correspond to the transmission sequence:

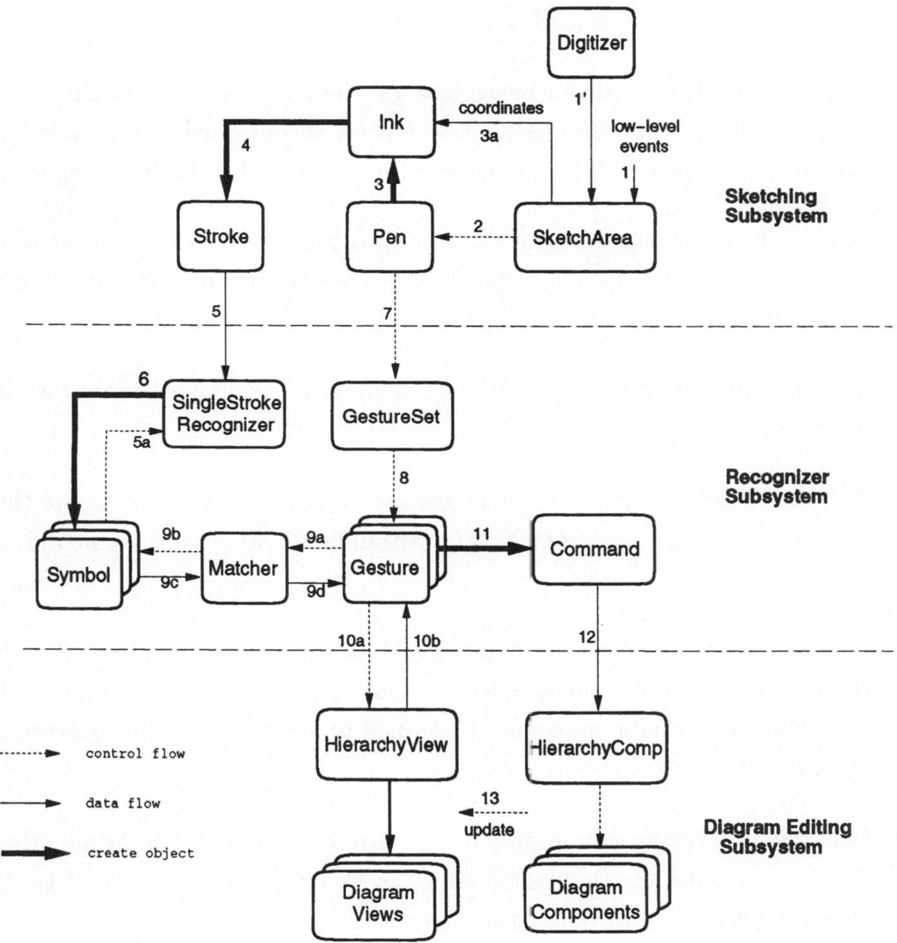

Figure 5.3: Overview of the most important Handi objects

1. The sketch area receives a pen-down event. In case that a digitizer is used, such an event comes from the digitizer.

2. The pen object is activated by the sketch area.

3. The pen creates a new ink object which performs inking and recording coordinates of pen-movements.

4. After receiving the pen-up event, the ink object creates a stroke object which encapsulates all the recorded coordinates.

5. Each stroke object will be recognized by the single stroke recognizer. As discussed in chapter 3, the recognition process is controlled by the underlying symbol hierarchy, which is illustrated with the arrow-line **5a** in the figure.

6. The result of the single stroke recognition is represented by an object of the class `Symbol`. This object is stored in the database and the incremental merging is activated.

7. The pen object starts the high-level recognition by using methods of the class `GestureSet`.

8. The `GestureSet` contains a list of gesture objects which performs the three basic recognition steps: matching, constraints checking, and command creation.

9. Each gesture object matches its own graphical symbol with symbols stored in the database by using the selective matcher of the low-level recognizer (**9a, 9b**). Matching-results are symbols returned to the corresponding gesture object (**9c, 9d**).

10. Constraints checking is supported by the corresponding views of the underlying diagram. **10b** depicts that useful data can be returned to the actual gesture object for dealing with the gesture semantics.

11. In case that the gesture constraints are fulfilled, a command object is created as the result of a recognition cycle.

12. Structure editing commands are transferred to appropriate diagram components for interpretation.

13. Manipulations of diagram components invoke updates of the corresponding graphical views.

5.1.2.3 Notations

Figure 5.3 gives an informal overview of the key abstractions and the global mechanisms of the Handi architecture. However, it is impossible to capture all the subtle details of a complex software system in such a figure. Skimming through the object-oriented design methods, one can find that many graphical notations in form of diagrams are developed to support object-oriented analysis and design. In the following sections, we use the object diagram and class diagram introduced by Booch [11] to describe the software architecture of Handi. Figure 5.4 shows the components used in this notation.   A class diagram is used to show the existence of

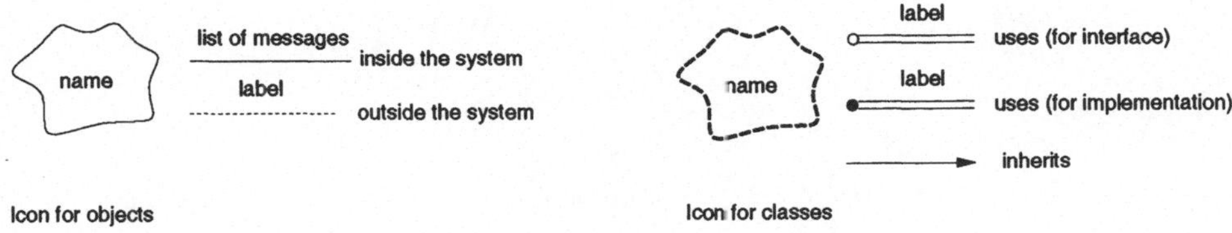

Figure 5.4: Booch's notations

classes and their relationships in the logical design of a system. One of the most important advantages of Booch's class diagram is the possibility to represent various relationships between classes rather than just the inheritance relationship such as in traditional class trees. An object diagram can be used to show the existence of objects and their relationships that illustrate the semantics of key mechanisms. Classes are largely static in the design of a system, whereas objects are much more transitory, in that many of them may be created and destroyed during the execution of a single program.

5.2 Sketching Subsystem

A handsketch-based diagram editor supports a gestural input model. Handi-based editors use the sketching subsystem to allow the user to sketch diagrams with computers in the same style as with paper and pen. The sketching subsystem has the primary task to transform low-level input data such as digitizer data or mouse events into high-level stroke objects. The architecture defines classes and objects within this subsystem. Figure 5.5 presents the classes and their relationships. The object diagram 5.7 specifies how objects of this subsystem interact on each other and how they communicate with objects of other subsystems.

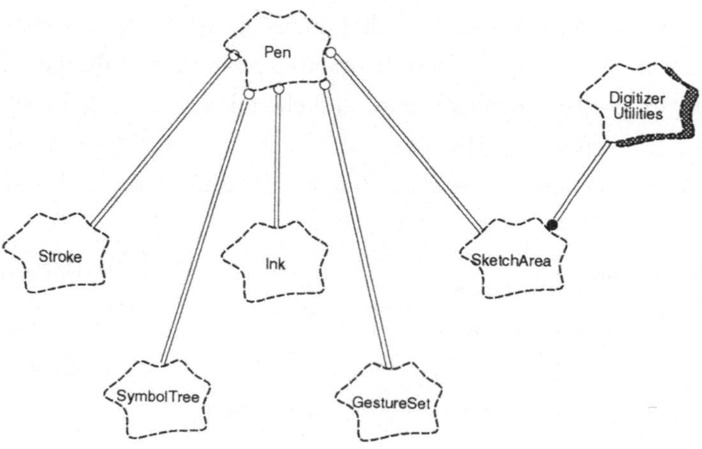

Figure 5.5: Class diagram of the sketching subsystem

5.2.1   Sketching Area

A sketching area processes user input events and displays graphical editing objects. A sketching area is therefore implemented in terms of window system or toolkit abstractions; the `SketchingArea` class might be derived from `Drawable` of the X Window System [36], `DrawingArea` Widget of the Motif Toolkit [94], or `Viewer` of Unidraw [129].

A sketching area object defines event handlers and direct manipulation tools which can be used with it. For drawing diagrams, a pen is the default tool. Both a mouse and a digitizer pen can be used as the input device for sketching diagrams. In case that a digitizer is used, the digitized absolute coordinates must be transformed to pixel coordinates compatible with the underlying window system.

As the research started, pen-based computers were not available. We had the idea to combine a graphical workstation with a sonic digitizer [109] so that the user

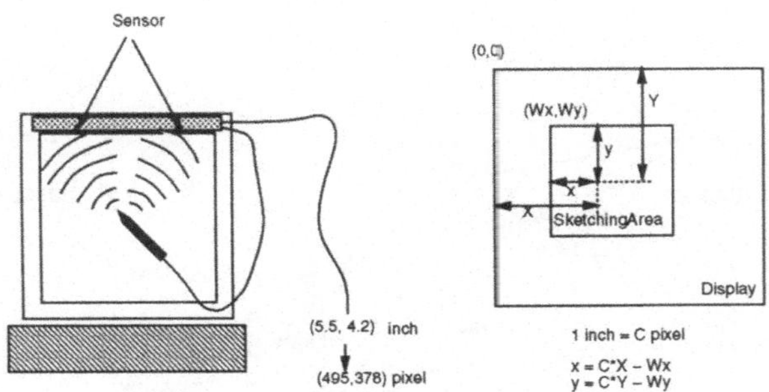

Figure 5.6: A sonic digitizer can be used as an input device.

can sketch with the digitizer pen on the workstation's display just like on paper [143]. A sonic digitizer uses sound waves to determine the distance between the position of the pen on the display and two microphone sensors, which can be mounted on the frame of the display without any problem. Figure 5.6 illustrates our experiments enabled us to use a conventional graphical workstation as a pen-based computer, as well as the schematic for how to transform absolute coordinates into window coordinates.

5.2.2 Pen

Pen is the key abstraction for the gestural input method. On the analogy of paper and pen, the **Pen** class encapsulates the functionality of a real pen, that is, writing and drawing. A pen object creates an ink object at the start of a sketch. Further, a pen object controls the gestural dialog, coordinates the low-level recognizer and the high-level recognizer. Figure 5.3 shows that the starting both of the low-level recognizer and the high-level recognizer is controlled by the pen object.

5.2.3 Ink

Myers [83] indicated that there are basically five interaction tasks: select, position, orient, path, and text. Under this classification, ink belongs to the path-interactor which is used to get all of the points a mouse or a digitizer pen goes through between pen-down and pen-up events. The design of the **Ink** class is based on

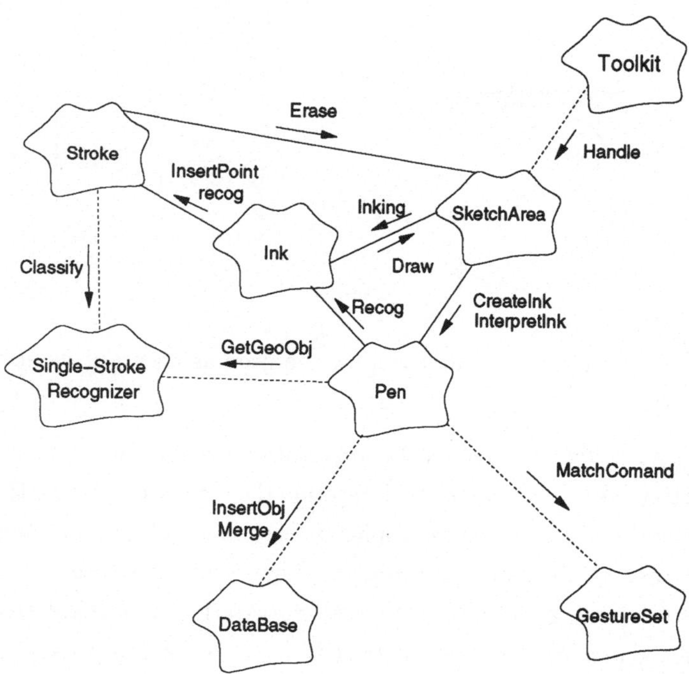

Figure 5.7: Object diagram of the sketching subsystem

the observation that gestural interfaces need different interaction techniques than
conventional interfaces. In contrast to rubberbanding or dragging, inking traces the
pen-movement by drawing iteratively the current pen positions.

5.2.4   Stroke

While the ink object concentrates on the handling of move-events and displaying
the current pen position, a stroke object is a container for encapsulating recorded
coordinate points. In other words, all point coordinates between a pen-down and
a pen-up event are represented in a single stroke object. Further, the class **Stroke**
defines basic protocols for using the single-stroke recognizer to transform a rough
stroke into a regular graphical object, and to remove itself from the display.

5.3 Recognizing Subsystem

Chapter 3 and 4 presented our concept for the low-level and high-level recognition, respectively. In this section, the software architecture, which integrates both the low-level recognizer and the high-level recognizer, is now described. Figure 5.8 captures the class structure of the recognizing subsystem. Here we find both interface relationships and inheritance relationships between classes. The key abstractions of the low-level recognizer are `SymbolTree` and `Symbol`, respectively. The key abstractions of the high-level recognizer are `GestureSet` and `Gesture`. The class `SymbolTree` builds the interface to the low-level recognizer. Both `Symbol` and `Gesture` are designed as base classes for subclassing. While the class hierarchy rooted by the class `Symbol` corresponds to the symbol hierarchy designed in chapter 3 (figure 3.9); subclasses of the class `Gesture` are editing gestures. Handi provides several frequently used gestures for direct use. Usually, only subclasses of `CreateGesture` must be specified by the editor developer for each individual diagram language.

5.3.1 Symbol

`Symbol` is the base class which realizes the object-oriented, hierarchical recognition method that is discussed in chapter 3. The hierarchy rooted by `Symbol` is clearly a specialization-hierarchy. Each node in the hierarchy represents a symbol class. Superclasses represent generalized abstractions and subclasses represent specializations. Such hierarchy is also called *kind-of* hierarchy as described in [11]. For example, square is a *kind of* rectangle, rectangle is a *kind of* parallelogram.

The basic idea of our low-level recognition is that *objects recognize itself*. This idea is represented in the Handi architecture by the basic protocol `Specialize` for the root class `Symbol`. Each class in the class-hierarchy provides an appropriate method according to this protocol for recognizing an object of this class into an object of its subclass. Drawing and erasing graphical objects build the additional protocols.

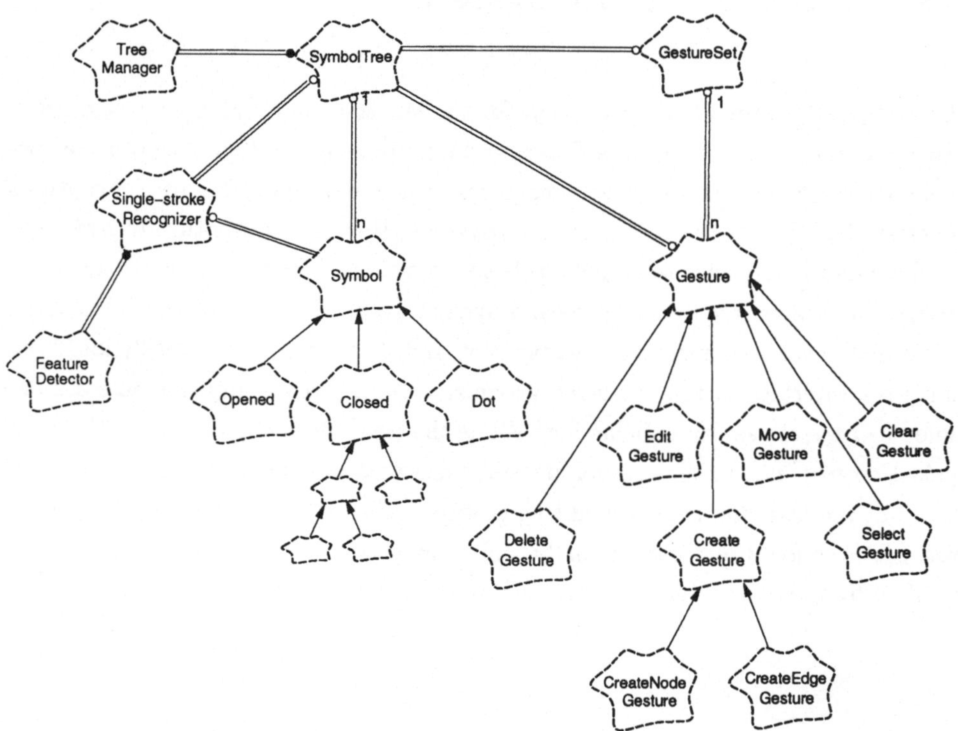

Figure 5.8: Class diagram of the recognizing subsystem

5.3.2 SymbolTree

The SymbolTree is primarily a container of graphical objects which are instances
of the class Symbol. This class realizes the concept of the hierarchical database
discussed in chapter 3. An object of the class **TreeManager** supports the hierarchical organization, and provides methods to access symbols stored in a symbol tree.
Messages such as inserting a new symbol into the database and merging connected
symbols within the database, can be passed to a symbol tree. An object of the class
SymbolTree provides interfaces for the high-level recognizer to the low-level recognizer. The main interface protocols are GetSymbol for matching gesture shapes and
RemoveSymbol for removing symbols in the database.

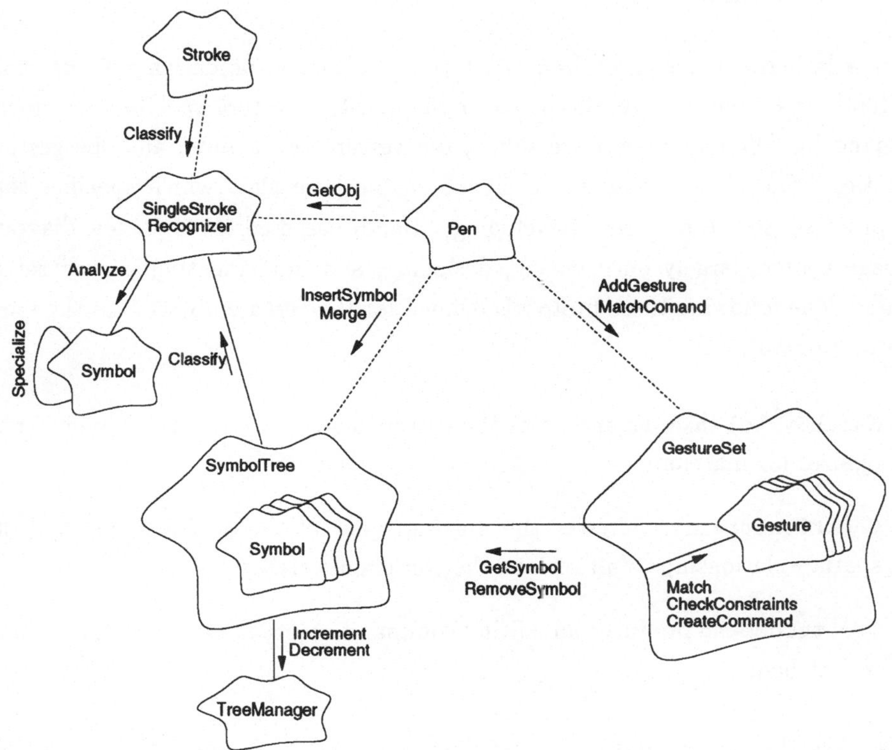

Figure 5.9: Object diagram of the recognizing subsystem

5.3.3 Single-Stroke Recognizer

The class `SingleStrokeRecognizer` is designed as an interface class to access our object-oriented recognition method presented in section 3.4.3. An instance of this class manages the iterative recognition loop by creating and destroying objects of the class `Symbol` or its subclasses.

A single-stroke recognizer can be invoked both by a stroke object of the sketching subsystem and by the `SymbolTree` object. A stroke object uses the single-stroke recognizer for transforming a rough stroke into a graphical symbol. The `SymbolTree` uses the single-stroke recognizer for recognizing a merged symbol in the incremental update process. In this case, a merged symbol is sent to the single-stroke recognizer via the `Classify` operation as shown in figure 5.9.

5.3.4 Gesture

`Gesture` is perhaps the most important class for an editor designer in developing Handi-based editors. As discussed in chapter 4, a gesture specifies an editing command by defining the gesture shape, the gesture constraints, and the gesture semantics. The class `Gesture` is designed as the base class which specifies top-level protocols for all gestures. Developing a Handi-based editor for a new diagram language centers largely on choosing, designing, and implementing a good set of gestures. The following three protocols defined for the gesture objects are the most important operations:

- `MatchSymbol` takes as argument the gesture shape and uses the low-level recognizer for matching.

- `CheckConstraints` accesses the diagram view of the actual diagram, tests spatial relationships, and returns the constraints states.

- `CreateCommand` produces an editing command object in case that the gesture is matched.

To allow the incremental parsing of diagrams, the constraints checker returns three different states for the basic protocol `CheckConstraints`. Figure 5.10 illustrates this with a Petri net example.

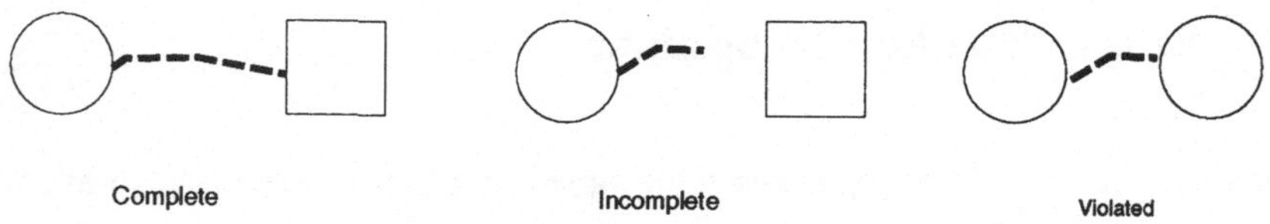

Figure 5.10: There are three different results by checking gesture constraints.

Complete means that the constraints are fulfilled.

Incomplete means that the constraints are not fulfilled, but the gesture produces no syntactical or semantical errors. This gesture is therefore incomplete, and the user can still complete it incrementally.

Violated means that the gesture is wrong. The user should be warned by an error message, and the gesture will be rejected. This is shown at the right hand side of figure 5.10, the constraints checker of the "create arc" gesture returns "violated" if the user tries to connect two places in a Petri net editor.

As shown in figure 5.8, Handi provides several frequently used gestures as system classes for direct reuse. These include the classes `CreateGesture`, `DeleteGesture`, `SelectGesture`, `MoveGesture`, `NameGesture`, and `EditGesture`. Moreover, the `CreateGesture` is subclassed to `CreateNodeGesture` and `CreateEdgeGesture` to encapsulate further common attributes. For example, the constraints defined for a `CreateEdgeGesture` are that the two endpoints of the gesture are connected with some connectors of node components. One of Handi's design goals is to encapsulate such common features into reusable base classes. For example, a Petri net editor can subclass the `CreateEdgeGesture` to define a new class `CreateArcGesture`. The constraints checker provided by the `CreateEdgeGesture` can be reused, one need only to add Petri net specific semantics into its constraints checker, such as a place can only be connected with transitions.

5.3.5 GestureSet

An object of the class `GestureSet` contains all gesture objects which are used for editing the underlying diagrams. The class `GestureSet` builds the interface to invoke the high-level recognition by a `Pen` object. `GestureSet` has access to the low-level recognizer via a `SymbolTree` object. The class `GestureSet` provides mainly two basic protocols:

- When constructing a pen object, all gesture objects are inserted into a `Gesture-Set` object via the operation `AddGesture`. This is similar to adding menu items in a menu bar.

- The operation `MatchCommand` starts the high-level recognition to carry out an editing command object. A `GestureSet` object has the functionality of a knowledge base which contains the diagram syntax in form of gesture objects.

5.4   Editing Subsystem

The design goals of the editing subsystem are on the one hand to span the gap between general editor frameworks and the implementation requirements of diagram editors, and on the other hand to provide spatial parsing utilities for the high-level recognizer which requires syntactical and semantic knowledge about the underlying diagram.

A diagram editor differs from general graphical editors mainly in that a diagram editor creates and manipulates structures rather than unstructured picture elements. Further, we have discussed in chapter 4 that the HiNet diagram languages have mainly two basic types of structures: hierarchy and connectivity. This is the fundamental starting-point of the basic abstractions of the editing subsystem of Handi. Indeed, the key abstractions of the editing subsystem are classes which are designed to represent and to manipulate these structures. Figure 5.11 is the class diagram of the editing subsystem which gives an overview of the key abstractions.

One design decision of the editing subsystem is to benefit editor functionality from general editor frameworks. Functionalities such as structure preserving editing like zooming and scrolling are therefore not in the scope of the Handi architecture. Saving and loading objects, as well as implementations of drawing and direct manipulation of graphical components are also facilities from general editor frameworks. A specific implementation of Handi establishes an interface between Handi and the used editor framework.

5.4.1   Hierarchy Component

General editor frameworks such as Unidraw [129] provide basic abstractions for graphical components which use graphic objects in both their subjects and views to define their appearances. Further, composite graphical components can define their appearances by assembling their subcomponents' graphics into composite graphics. However, composite graphical components which are defined in this way can only be considered as a single component. This has the consequency that hierarchical structures cannot be *directly* manipulated, because subcomponents are not accessible. Therefore, composite components of general editor frameworks cannot be used directly as hierarchy components to represent hierarchies within diagrams.

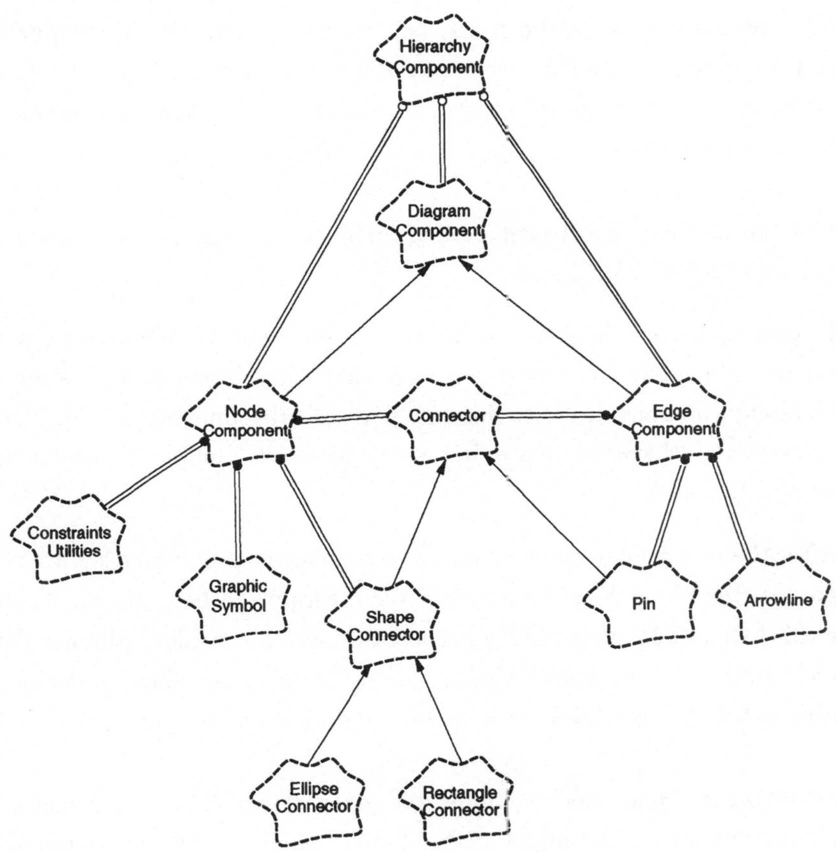

Figure 5.11: Class diagram of the editing subsystem

The class **HierarchyComponent** of the editing subsystem is designed to allow accesses to subcomponents. Objects of the class **HierarchyComponent** are usually not explicit *visible*, instead, an editor manages a top-level hierarchy component, and all other hierarchy components are attached to a corresponding node which manages other subnodes. This representation form of hierarchy is refined into an internal representation schema for the benefits of the high-level recognition which is discussed in section 4.3.1. A hierarchy component object is similar to a container object which manages a list of diagram components which are subcomponents of this object. The basic protocols of the class **HierarchyComponent** consider inserting and removing of subcomponents.

Within a HiNet diagram, a hierarchy component *cannot* be created directly. Instead, a hierarchy component can only be inserted implicitly by creating nodes. In chapter 4, we have presented the schematic of when and how a hierarchy component can be created. A "create node" command is interpreted by the top-level hierarchy component as follows:

1. Use the method `SmallestNodeEnclosing` to find the smallest node which encloses the considered node.

2. In case that no node is found, insert this new node as subcomponent of the top-level hierarchy component; in case that the selected node has already a hierarchy component, insert this new node as subcomponent of this hierarchy component, otherwise, create first a new hierarchy component as subcomponent of this node.

3. After the new node is inserted as a subcomponent of a proper hierarchy component, the hierarchical structure of the diagram can be changed. Therefore, a third step `UpdateHierarchy` is necessary, which considers whether the new node contains some other existing nodes. In this case, these nodes must be *reparented*, that is, change their parent hierarchy component.

`RemoveHierarchy` is another important protocol which supports hierarchical structure manipulation like undoing the "create node" command or removing a node but not deleting the subnodes of the node. The operation `RemoveHierarchy` is usually interpreted by the parent of the hierarchy component which should be removed. All subnodes of this hierarchy component will be considered as subnodes of its parent.

5.4.2　Hierarchy View

Editor frameworks provide abstractions for graphical components which use graphic objects in both their subjects and views to define their appearances. Object-oriented editor frameworks, which benefit by the Smalltalk's MVC mechanism [65], define usually separate communication protocols for component subjects and views. Handi is built on top of such a framework and defines separate protocols for hierarchy components and hierarchy views as well. As aforementioned, constraints in the high-level

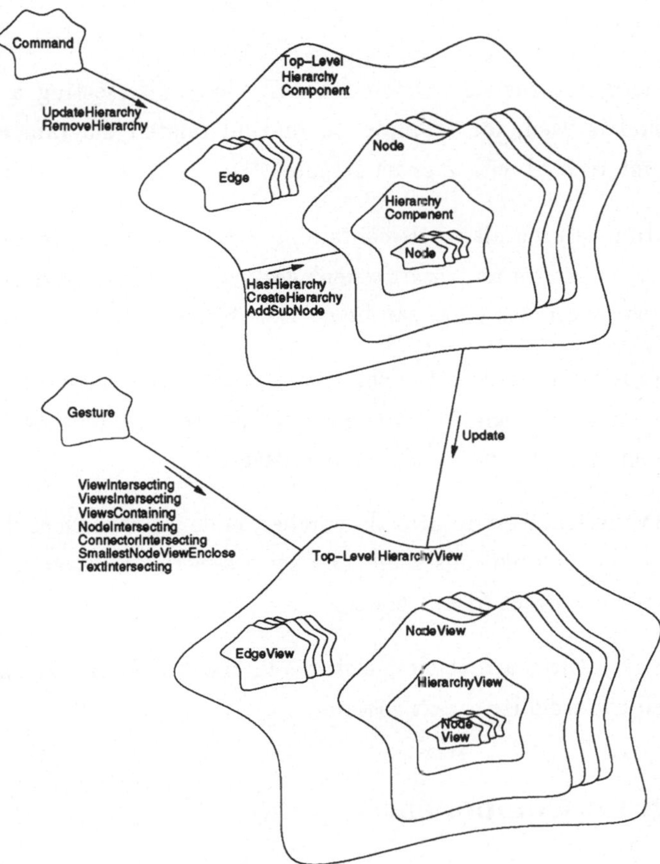

Figure 5.12: Object diagram of the editing subsystem

recognition are spatial relations between gestures and the external representation of the underlying diagrams. `HierarchyView` is the class which provides most of the spatial parsing utilities for the high-level recognizer. Figure 5.12 shows that a gesture object can pass the following messages to consider spatial relationships between a gesture symbol and the diagram view:

ViewIntersecting returns a subcomponent view intersecting a given box. This method is used for checking constraints such as a diagram component must not intersect any other components.

NodeIntersecting is the same as `ViewIntersecting`, however, only node views
are considered.

ViewsIntersecting returns all subcomponents' views intersecting a given box.
This method is used, for example, to find all related diagram components
which are intersected with a given gesture.

ViewsContaining returns all views containing a given point. For example, the
constraints, defined for an "insert token" gesture that the gesture shape should
be contained by a place view, can be checked by using this method.

ConnectorIntersecting returns the child `ConnectorView` that intersects a given
rectangular area. Recognition of connectivities uses this method to find connectors which are touched by a given handsketch.

SmallestNodeViewEnclose returns the smallest node view which encloses a given
box. A hierarchy related constraints checker passes this message to a hierarchy
view to find the correct parent node.

TextIntersecting returns a text view which is intersected with a given box. This
can be used to recognize a text gesture.

5.4.3 Diagram Component

`DiagramComponent` is the base class which represents all visible diagram components. One common characteristic of diagram elements is that each diagram component consists of a graphical symbol and a corresponding label which has a fixed
position relatively to the graphical symbol. For example, a state object has a name
which can be positioned at the upper-left corner of the graphical symbol. The class
`DiagramComponent` provides abstractions for defining the graphical symbol, the textual label, and their relative positions. Utilities for accessing this information are
defined for objects of the class `DiagramComponent`:

- `SetGraphic` sets the graphic of the diagram component.

- `SetText` sets the character string to the given one.

- `GetGraphic` returns the graphic of the diagram component.

- **GetText** returns the text of the diagram component.

Considering the graphical syntax and the structural information, diagram components can be further classified into node components and edge components as shown in figure 5.11.

5.4.3.1   Node

The class **Node** is an abstraction of diagram elements which represent *nodes*. Usually, nodes are graphically represented by a closed geometrical symbol. As discussed in chapter 4, nodes are used both for representing hierarchy and connectivity information. The hierarchical structure is represented visually in the form that nodes *contain* spatially subnodes. This hierarchical structure is represented internally by using hierarchy components as designed in section 4.1.1 and section 4.3.1. A connection is represented by an edge component and two connected nodes. Internally, each node manages four lists, each of which has utilities for adding and removing objects into and from the corresponding list:

1. A list of all nodes which are connected **from** this node can be manipulated by the methods **AddOutNode** and **RemoveOutNode**.

2. A list of all nodes which are connected **to** this node can be operated by the methods **AddInNode** and **RemoveInNode**.

3. A list of all edges which go out **from** this node can be accessed by the methods **AddOutEdge** and **RemoveOutEdge**.

4. A list of all edges which point **to** this node can be manipulated by the methods **AddInEdge** and **RemoveInEdge**.

Further, the class **Node** defines basic protocols for interpreting structure manipulation commands. Considering the connectivity, if a node should be deleted, all edges, which are connected to and from this node, are deleted automatically for maintaining consistency. The above four protocols allow easy accesses to the connection information. Considering the hierarchy, creating a node can implicate incrementing the hierarchy levels. A node can be passed an operation **AddSubNode**

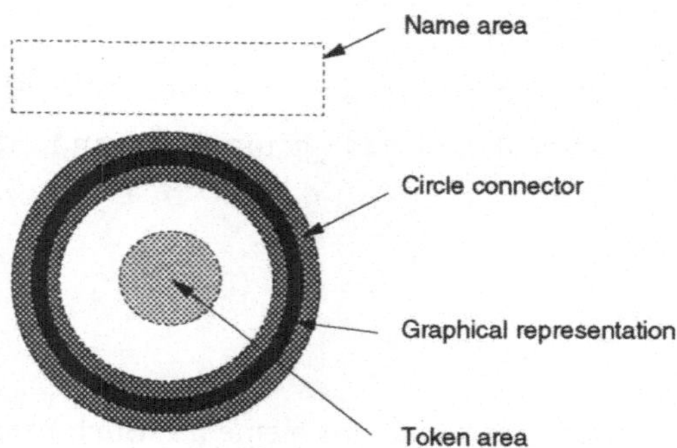

Figure 5.13: The composition of place component used in Petri nets

to insert a node as its subnode. The method `HasHierarchy` and `CreateHierarchy` are utilities to examine and to create a hierarchy component of this node.

The class `Node` is designed for subclassing to construct new language-dependent diagram elements. Standard composition of a node component consists of a closed graphical symbol, a label, and a shape connector for supporting connectivity. Subclasses of `Node` define additional subcomponents which realize application-dependent syntax and semantics. Figure 5.13 illustrates the design of the class `Place` which is a subclass of `Node` for the Petri net editor which will be described in chapter 7. The graphical symbol is a circle, the name label is aligned to top of the graphical symbol. Further, the class `Place` defines graphical components inside of the circle for representing tokens and a circle connector for maintaining connectivities.

5.4.3.2 Edge

The class `Edge` abstracts graphical components which represent the connective structures together with node components. An edge has conceptual four subcomponents: a graphical symbol, a textual label, and two pins for supporting connectivity. Internally, an edge manages two references, one for the node which is connected **to** this edge, and the other for the node which is connected **from** this edge. The graphical symbol can be a spline, a straight line, or a polyline as shown in figure 4.4.

While the connective structures are stored in nodes (four lists of references), the construction of a connection is initialized by creating edge objects. Simply speaking, when creating an edge object, appropriate messages are passed to connected nodes via the operations **AddOutNode**, **AddInNode**, **AddOutEdge**, and **AddInEdge**. On the contrary, when deleting an edge, the corresponding connection is destroyed by passing messages to the connected nodes via operations **RemoveOutNode**, **RemoveInNode**, **RemoveOutEdge**, and **RemoveInEdge**. Figure 5.14 is the object diagram to illustrate the protocols defined for constructing and destroying connections.

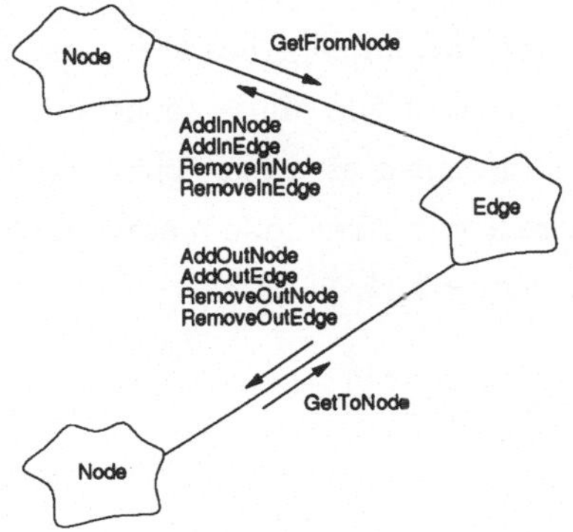

Figure 5.14: Protocols defined for node and edge objects

A connection is represented graphically by an edge symbol and the spatial coincident relationships between the endpoints of this edge and the connected nodes. Within the design of Handi, nodes can be moved by using direct manipulation techniques such as dragging. The class **EdgeComponent** is responsible to satisfy connection constraints after the connected nodes are moved. For this reason, the class **EdgeComponent** defines the method **Update** to be invoked after each manipulation of connected nodes.

5.4.4 Connector

Editor frameworks such as Unidraw support connectivity and confinement semantics by providing abstractions in form of connectors. Connected connectors can affect each other's position in specific ways defined by connector's mobilities. Unidraw supports, for example, *pin*, *slot*, and *pad* connectors. While the connection semantics of edges can be realized by using the Unidraw's pin connector, the connection semantics of nodes require further considerations. This is because the connection semantics of nodes are defined that the endpoints of an edge must stay continually on the boundary of a node's graphical symbol. Therefore, Handi defines the class `ShapeConnector` which can be a subclass of editor framework's `Connector` or a root class dependent on its implementation.

The connection semantics of `ShapeConnector` are defined as follows: A shape connector can only be connected with a pin for representing connective structures. A node carries a shape connector, an edge has two pins. If a pin is connected with a shape connector, the position of the pin is constrained on the boundary of the shape. `ShapeConnector` is designed as a base class which must be subclassed for each concrete shape. Currently, a `CircleConnector` and a `RectangleConnector` are developed as shown in figure 5.11.

5.5 Summary

This chapter describes the Handi architecture by classifying Handi into three subsystems: sketching, recognizing, and editing. An editor for a particular diagram language relies on the sketching subsystem for handling handsketch-based input, on the recognizing subsystem for gesture recognition, and on the editing subsystem for its structure representing and editing capabilities. The basic idea of Handi is to encapsulate common characteristics of handsketch-based diagram editors into classes by using object-oriented methodology. While many classes can be used directly for constructing handsketch-based editors, several classes are designed as *base* classes which must be subclassed for adding language-specific syntax and semantics.

Conceptually, Handi is independent of any editor framework. However, a concrete implementation of Handi *uses* a concrete editor framework because Handi is designed on the top of general editor frameworks whose functionality should be reused as much as possible. Therefore, the overall structure of a concrete Handi application depends partially on the used editor framework. In the next chapter, we describe a prototype implementation of Handi which is based on the Unidraw architecture.

Chapter 6

Implementation

In this chapter we focus on the salient aspects of our Handi implementation. We consider the overall structure of the system, the key classes, relevant algorithms, and novel techniques we applied. The current Handi implementation has basically two parts: 1) the recognizing subsystem is implemented in reusable C++ classes independent of any editor framework or window system; 2) the sketching and editing subsystems are implemented on top of the editor framework Unidraw. First we give an overview of our implementation strategy, the object-oriented programming language we chosen, and the history of how this implementation originates. Then the significant aspects of the implementation are described in two separate sections.

6.1   Overview

It should be mentioned that the implementation of Handi is *not* done in one step after the Handi architecture was designed completely. In contrast, as indicated by Gall [34]: "A complex system that works is invariably found to have evolved from a simple system that worked... A complex system designed from scratch never works and cannot be patched up to make it work. You have to start over, beginning with a working simple system." Indeed, the design of the Handi architecture, its implementation, and the experimental applications have been accompanied from the beginning. The current Handi implementation has been gained in an iterative, incremental process which received permanently input both from the improvements of the Handi architecture and from new requirements obtained by prototyping experimental applications.

Object-Oriented Programming and C++

Until 1990, C was the only programming language used at Cadlab[1]. Because our software developments are based on the X Window System [36] and the X Toolkit Intrinsics [79], we are familiar with the object-oriented programming style used in the implementation of the X Toolkit. Most code of the early versions of Handi were implemented in C by using this object-oriented programming style.

After C++ was available at Cadlab in 1990, the author switched to C++ with great interest in object-oriented programming by using an object-oriented language. All existing C code was transformed in a limited time to C++. C++ is an attractive language because it supports object-oriented programming without compromising the execution efficiency of C. While C++ lacks useful features such as run-time access to class information, untyped message passing, and garbage collection, its explicit object-oriented semantics greatly simplifies the implementation of the object-oriented Handi architecture compared to C.

Our Handi prototype consists of three C++ libraries containing about 13,000 lines of source. It runs on top of Unidraw and the X Window System. Table 6.1 presents a breakdown of the prototype implementation in classes and lines of source code.

Table 6.1: Handi prototype libraries code breakdown

Subsystem	Classes (No.)	Code (lines)
Sketching	5	467
Recognizing	50	5127
Editing	29	7019

The recognizing subsystem is implemented independently of any editor framework; the sketching and editing subsystems are implemented directly on top of Unidraw by subclassing proper Unidraw classes.

---

[1]Cadlab is a joint venture of Paderborn University and Siemens Nixdorf Informationssysteme AG. The author joined Cadlab 1988, and has been working from the beginning in the user interface group.

Table 6.2: Classes of the recognizing subsystem

Base class	Derived class					
GestureSet						
Gesture	DeleteGesture					
	EditGesture					
	NameGesture					
	MoveGesture					
	ClearGesture					
	SelectGesture					
	CreateGesture	NodeGesture				
		EdgeGesture				
	CharGesture					
Symbol	Closed	Polygon	Triangle			
			Quadrilateral	Trapezoid		
				Parallelogram	Rectangle	Square
		Ellipse	Circle			
	Opened	Line	VerticalLine			
			HorizontalLine			
		MultiLine	Arrow	RightAngle	LForm	
				SharpArrow		
			TriLine	UForm		
				ZForm		
				Bottle		
				Basin		
		Arc				
	Dot					
	Composite					
SSRecognizer SymbolTree TreeManager						
Angle FCode						
Detector	CornerDetector					
	LineDetector					
	ArcDetector					
CharRecognizer						

6.2 Recognizing Subsystem

Table 6.2 gives an overview of classes designed for the recognizing subsystem which is implemented as an extensible C++ class library. The recognizing subsystem of

Handi consists of a high-level recognizer and a low-level recognizer. The high-level recognizer is implemented mainly by the two base classes `GestureSet` and `Gesture`. The implementation of the low-level recognizer consists of eight base classes. The class hierarchy rooted by the class `Symbol` implements the main part of the hierarchical recognition algorithm discussed in chapter 3. The class `SSRecognizer` is an interface class which encapsulates the control structure of the hierarchical recognition process.

GestureSet and Gesture

```
class GestureSet {
public:
   GestureSet(Editor*);

   void Add(Gesture*);
   void Remove(Gesture*);
   Command* MatchCommand(SymbolTree*);
private:
   UList   _gestures;
   Editor* _editor;
};
```

A `GestureSet` object maintains a list of gesture objects which can be inserted and removed by `Add` and `Remove`. The method `MatchCommand` invokes the gesture recognition by delegating the recognition task to each responsible gesture object which is defined as follows:

```
typedef Symbol* (*SelectiveMatchFunc)(SymbolTree*);
enum ConstraintState {Complete, Incomplete, Violated};

class Gesture {
public:
   Gesture(SymbolType, SelectiveMatchFunc);

   boolean Match(Editor*, SymbolTree*);
   virtual Command* CreateCommand(Editor*);
   Symbol* GetMatchResult();
protected:
   virtual ConstraintState CheckConstraints(Editor*);
protected:
```

```
    SymbolType          _gesture_shape;
    Symbol*             _match_result;
    SelectiveMatchFunc _selective_match;
};
```

A `Gesture` object specifies an editing gesture by defining the gesture shape, the gesture constraints, and the gesture semantics as described in chapter 4. While the gesture shape can be configured by creating a gesture object, gesture constraints and gesture semantics are specified by subclassing this base class and implementing proper `CheckConstraints` and `CreateCommand` methods. Currently, ten frequently used gestures are implemented by subclassing the base class `Gesture` as shown in table 6.2. The programming efforts which the application programmer must invest are therefore very limited.

Symbol Hierarchy

This section describes the implementation of the symbol hierarchy by giving relevant C++ code fragments of representative classes.

```
class Symbol {
public:
    Symbol(int *x, int *y, int n);
    virtual SymbolId Specialize(Symbol*& result);
protected:
    int *_x, *_y;
    int _count;
};
```

Each handsketch, which is represented by a sequence of point coordinates in a stroke object, can be transformed into a `symbol` object. The protected members `_x`, `_y`, and `_count`, which are the constructor parameters of this class, store the information of a handdrawn stroke.

All subclasses of `Symbol` have similar declarations. The declarations of class `Opened` and class `Line` are selected to give two representative examples as follows:

```
class Opened : public Symbol {
public:
    Opened(int *x, int *y, int n);
```

```
    virtual SymbolId Specialize(Symbol*& result);
};
```

This class declaration means that the class **Opened** inherits the structure of **Symbol**, i.e. the class **Opened** is just like its superclass **Symbol** except that an object of this specialized class has a different constructor and a different implementation of virtual member function **Specialize()**. The members _x, _y, and _count which are used to store the point coordinates are inherited from **Symbol**. Only single inheritance is used in the class hierarchy rooted by **Symbol**.

```
class Line : public Opened {
public:
    Line(int x0, int y0, int x1, int y1);
    virtual SymbolId Specialize(Symbol*& result);
private:
    int        _length;
    Angle      _slope;
};
```

The class **Line** is a subclass of **Opened**. It inherits members of its superclasses and has two additional members to store "specific" properties, the length and the slope of a line object, which **Opened** objects do not have.

Besides the constructors, the key member function of all classes in this class hierarchy is the member function **Specialize()**. This function returns the new class id which is more specific than **this** object. Additionally, a new object is returned via a reference parameter. If a specialization is not possible, **this** object is returned. To give an idea how such a function can be implemented, the code of a simple member function of the class **Symbol** is appended below.

```
SymbolId Symbol::Specialize(Symbol*& result) {
   if (_count == 1) {
      delete result;
      result = new Dot(_x[0], _y[0]);
      return DOT;
   } else {
      /* distance between first and last point */
      int d = Distance(_x[0], _y[0], _x[_count-1], _y[_count-1]);

      /* thresholds are dependent on the dimension of object */
```

```
        if (d < threshold_closed) {
            delete result;
            result = new Closed(_x, _y, _count - 1);
            return CLOSED;
        } else if (d > threshold_opened) {
            delete result;
            result = new Opened(_x, _y, _count);
            return OPENED;
        } else {
            result = this;
            return SYMBOL;
        }
    }
}
```

Most of the `Specialize` member functions are not so trivial as this one of the class `Symbol`. Several feature detectors are required to make local decisions. These detectors are encapsulated into appropriate classes.

Single-Stroke Recognizer

The control process of the single-stroke recognizer is encapsulated into the class `SSRecognizer`. The method `Classify` invokes the following iterative recognition process as discussed in chapter 3:

```
SymbolId old_type, this_type = SYMBOL;
symbol = new Symbol(x, y, npoints);
do {
    old_type = this_type;
    this_type = symbol->Specialize(symbol);
} while (old_type != this_type);

result = symbol;
```

The point coordinates (`x, y, npoints`) captured from the input device are passed from a stroke object to the constructor of the class `Symbol`. An object of the class `Symbol` is created. The recognition process is controlled within a loop by 'specializing' repetitively the object `symbol` to a new one. After completing the while-loop, the object `symbol` contains the recognition result. The essential point is the use of polymorphism which is one of the most powerful features of object-oriented programming languages such as C++. This feature distinguishes the object-oriented programming from the traditional programming with abstract data types. In the above implementation, the variable `symbol` is defined as a pointer to an object of the root class `Symbol*`, which can point to objects of any its subclasses at the run time. Using later binding by declaring the member function `Specialize()` as virtual, this function becomes to be polymorphic.

The advantage of object-oriented programming is shown here clearly by means of the realization of the controlling structure of the recognition process. The complicated problem of handdrawn figure recognition is solved in a straightforward fashion. Specially the recognition process is controlled automatically.

Feature Detectors

Several feature analyzers are needed in the recognition process. These feature analyzers are used in the `Specialize` member functions as shown above. The most important feature analyzers are all encapsulated into classes rooted by the class `Detector`. The current implementation of Handi consists of mainly three feature detector classes: `CornerDetector`, `LineDetector`, and `ArcDetector`.

Algorithms of feature analyzers, which are described in chapter 3, are implemented as member functions of each class. To check whether an object owns a specific feature, an object of a desired feature-analyzer can be created. Calling the desired member function of the feature analyzer, appropriate features will be returned. For example, to check whether a `Closed`-object can be specialized to a `Polygon`-object, a corner detector is used to find all corners of the object and a line-verifier is needed to ensure that all substrokes between two adjacent corners are lines.

The declaration of the `CornerDetector` class looks like this:

```
class Angle;
class FCode;

class CornerDetector : public FeatureAnalyzer {
public:
    CornerDetector(int *x, int *y, int npoint);

    int GetCorners(int*&);
```

```
private:
    Angle *_angles;
    FCode *_fcodes;
    int   *_corners_idx;
    int    _corners_count;
};
```

Utility Classes

A class `Angle` and a class `FCode` are designed to simplify the frequently used calculations with angle values and Freeman's codes [32]. Both classes are used for members of the class `CornerDetector`. The design and implementation of these classes are based on abstraction and encapsulation in a standard way. Many useful functions for calculations with angle values and Freeman's code [32] are implemented as member functions of these classes. Freeman's code is useful for recognizing the drawing directions of a stroke. The class `LineDetector` uses the Freeman's code as well.

Character Recognition

One of the design goals of the Handi architecture is to allow easy integration of other recognition algorithms. For examining this design goal, an interface class `CharRecognizer` has been designed for encapsulating handwriting recognition algorithms. In the current implementation, a simple fuzzy-logic based algorithm [1] is integrated in the Handi system for entering short object names. The class `CharGesture` allows using handwritten characters as gestures for invoking editing commands.

6.3   Sketching and Editing Subsystem

Unidraw

As the research started, no editor framework was available. However, the design of the Handi architecture required iterative experiments and incremental refinements with running prototype editors. For this reason, the author began to build an editor framework, although, this is only peripherally related to the topic at hand, namely handsketch-based diagram editing. Some aspects of this unpublished editor framework can be found in [42].

After Unidraw [129] was available, we found that Unidraw provides exactly the basic functionalities which Handi requires. Using Unidraw, it is possible to concentrate on the main topic, and to be free from many time consuming implementation work for basic functionality such as store and load graphical components, connector semantics, zooming, scrolling, and many others. See the decision today, Unidraw builds a good base for developing Handi for the reason that diagram editors are specific object-oriented graphical editors.

Unidraw Abstractions

The basic abstractions of the Unidraw architecture are reflected in four class hierarchies:

1. **Components** represent the elements in a domain, encapsulate the appearance and semantics of these elements.

2. **Tools** support direct manipulation of components. Tools employ animation and other visual effects for immediate feedback to reinforce the user's perception that he is dealing with real objects.

3. **Commands** define operations on components and other objects. Commands are similar to messages in object-oriented systems in that components can receive and respond to them.

4. **External representations** convey domain-specific information outside the editor.

The Unidraw architecture provides base classes for component, command, tool, and external representation objects. Subclasses implement the behavior of their instances according to the semantics of the protocol defined by their base class. The partitioning of graphical object editor functionality into these four base classes is the foundation of the Unidraw architecture.

Unidraw Application

Based on the above key abstractions, Unidraw lays down the general structure of
Unidraw-based editors, which is reflected by the following objects:

- A **viewer** displays a graphical component view and provides an interface to
  scrolling and zooming it. Further, a viewer processes user's input events and
  translates them to conform to the Unidraw protocol.

- An **editor** provides a complete user interface to editing a graphical component
  subject. It unites one or more viewers with the commands and tools that act
  upon the component and its subcomponents.

- A **selection** object is a convenient interface to managing a set of distinguished
  graphical component views.

- A **catalog** provides independent name-to-object mappings for managing a
  database of components, commands, and tools.

- A **unidraw** object implements the main loop of the program, makes editors
  appear and disappear on the screen, maintains a log of all commands that
  have been executed and reverse-executed to support arbitrary-level undo and
  redo, and stores a reference to the catalog.

In terms of these objects, the general structure of Unidraw-editors has five levels:
At the highest level, a one-of-a-kind `unidraw` object communicates and coordinates
inter-editor operations. At the second level, an `editor` object associates tools and
user-accessible commands with one or more `viewers`. At the third level, each viewer
displays a graphical component view, most often the root view in a hierarchy. At
the fourth level, component views can be composed of other views to reflect their
subjects' structure. At the bottom level, component subjects can contain hierarchies
of subcomponent subjects to incorporate into a larger work.

Handi-based editors have similar general structures as Unidraw-based editors, because Handi uses extensively the general editing functionality provided by Unidraw.
The recognizing subsystem is integrated in each Handi application automatically by
using appropriate Handi classes which are subclasses of Unidraw classes. Therefore,
one of the most important issues of the Handi implementation is to select appropriate

Unidraw classes, and to build subclasses for realizing the Handi architecture. Table 6.3 shows the classes implemented for the sketching and the editing subsystems of Handi.

Table 6.3: Classes of the sketching and editing subsystems

Subsystem	Unidraw class	Handi Class		
Sketching	Tool DragManip Viewer GrowingVertices	Pen Ink SketchingArea Stroke		
Editing	GraphicComps	HierarchyComp DiagramComp	 NodeComp EdgeComp	
	GraphicViews	HierarchyView DiagramView	 NodeView EdgeView	
	DragManip	HierarchyDragManip		
	PinComp Connector	PinConnector ShapeConnector	 RectangleConnector EllipseConnector	
	Command	EnterNameCmd NetlistCmd PostScriptCmd		
	PreorderView	NetlistView	NLHierarchy NLDiagramComp	 NLNode NLEdge
	PostScriptViews	PSHierarchy PSDiagramComp	 PSNode PSEdge	
	Editor Creator Catalog	DiaEditor DiaCreator DiaCatalog		

Pen

The sketching subsystem consists of mainly four classes: Pen, Ink, SketchingArea, and Stroke. Perhaps the class Pen is the most important abstraction of the sketching subsystem. We implemented this class as a subclass of the Unidraw's abstraction

Tool, because the gestural interface is a kind of direct manipulation interfaces, and a pen is a tool for sketching and writing as well.

```
class Pen : public Tool {
public:
    Pen(SymbolTree*, GestureSet*);
    Manipulator* CreateManipulator(Viewer*,Event&,Transformer*);
    Command* InterpretManipulator(Manipulatcr*);
private:
    SymbolTree* _symbols;
    GestureSet* _gestures;
};
```

The constructor of Pen carries two objects which are stored into _symbols as the interface to the low-level recognizer, and into _gestures as the interface to the high-level recognizer. The basic protocol CreateManipulator defined for the base class Tool is used for creating a proper ink object, as well as the protocol InterpretManipulator for invoking the gesture recognition mechanism discussed before.

SketchingArea is implemented as a subclass of Viewer by installing a pen object as the current tool. Ink is derived from the Unidraw's class DragManip by defining its own new manipulating method to implement the ink semantics. Lastly, we subclass GrowingVertices to implement Stroke as a specific kind of rubberband.

Components

Vlissides [129] indicated that the main efforts for building Unidraw-based applications are choosing, designing, and implementing the required domain-specific components. This requires a comprehensive understanding of the Unidraw architecture and implementation details. One of our design goals of the Handi architecture is to provide as much as possible common attributes of diagram editors so that the application programmer is unburdened from many system details which are required for building diagram editors.

The Unidraw architecture supports compositions of graphical components by providing the class GraphicComps. The class DiagramComp inherits the class GraphicComps for reusing this composition functionality. Moreover, DiagramComp defines specific composition rules and utilities so that each object of this class consists of

a label and a graphical symbol. The class `DiagramComp` is further subclassed into
`NodeComp` and `EdgeComp` by implementing specific semantics as discussed in chapter
5. For example, `NodeComp` supports methods for building hierarchical structures and
for establishing connections with other edge components.

The implementation of `HierarchyComp` inherits the composition functionality of
`GraphicComps`. Further, hierarchy manipulation methods such as `UpdateHierarchy`
and `RemoveHierarchy` are implemented. One important extension to `GraphicComps`
is the spatial parsing functionality which is realized as a new command interpreter
for creation commands. If a hierarchy component interprets a creation command,
the new component is not inserted immediately as a direct subcomponent. Instead,
the object to be inserted is parsed by using the spatial containment relationship as
discussed in the previous chapters.

Views

One of the significant design issues of the editing subsystem is the representation
and manipulation of hierarchical structures. In contrast to Unidraw in which all
views composited in an object of the class `GraphicViews` are treated as a single
object, Handi allows the user to select and directly manipulate diagram components
which belongs to several different hierarchy levels. The main implementation effort related to this point was the reimplementation of all selection methods such
as `ViewIntersecting`, `ViewsIntersecting`, and `ViewsContaining`. Further, the
methods `SmallestNodeViewEnclose`, `IntersectedNodeView`, `NodeIntersecting`,
and `ConnectorIntersecting` were implemented to support the high-level recognizer for checking spatial relationships.

The `DiagramView` supports recognition of text gestures by providing the protocol `TextIntersecting` which detects whether a given box is intersected with the
label of a diagram object. Subclasses of `DiagramView` support more specific structure oriented direct manipulations. For example, the method `CreateManipulator`
returns an object of Handi's class `HierarchyDragManip` instead of Unidraw's class
`DragManip`. The `HierarchyDragManip` manipulator constraints the drag manipulation of a node object inside its parent node.

Connectors

Unidraw provides a good framework for building connectors. However, Unidraw's connection solver solves currently connector constraints independently in the horizontal and vertical axes. Thus, Unidraw cannot support, for example, non-linear connector glue and non-orthonormal slots and pads [128].

Within a HiNet diagram, the node-edge connection style is defined by constraining the endpoints of an edge object permanently on the boundary of the connected node object, which is obviously a non-linear connection. Instead of implementing a vector based `csolver`, our implementation strategy is to fix the positions of all nodes, tensing the two endpoints of each edge, and forcing the positions of the endpoints of edges on the boundaries of connected nodes during the update of each edge component.

Utilities

Within our Unidraw-based Handi implementation, several utility classes were implemented:

NetlistView supports an EDIF-like representation of HiNet diagrams. This opens the possibility to use Handi applications together with other tools.

PostScriptView provides PostScript representation of HiNet diagrams.

DiaEditor provides the predefined look and feel of a handsketch-based diagram editor.

DiaCatalog, DiaCreator support read and write of Handi components which are necessary for all Unidraw applications.

Figure 6.1 shows the look and feel of a Handi-based editor. A Handsketch-based diagram editor has usually much less command buttons than a diagram editor with conventional graphical user interface. The standard menu items for invoking file commands are always the same. Therefore, it is possible to encapsulate common user interface configurations into the class `DiaEditor` so that an application programmer

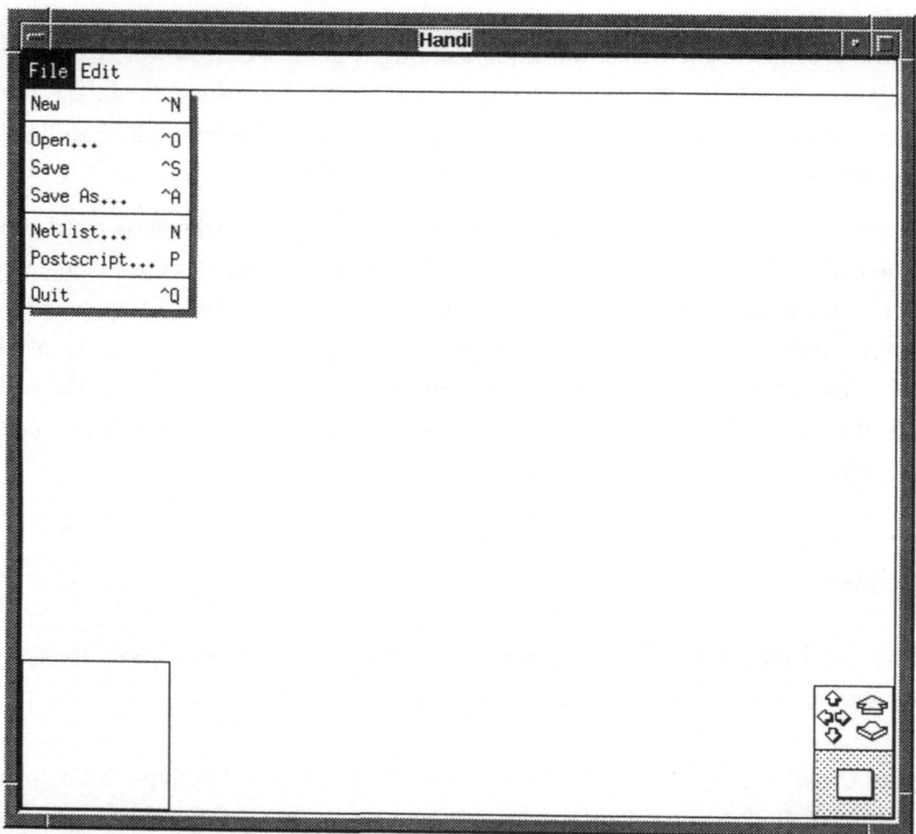

Figure 6.1: The look and feel of a Handi-based application

needs only to define language-specific gestures. The "Handi-standard" user interface components which are implemented in `DiaEditor` are:

1. Menus for standard file operations and the quit command.

2. A sketching area for drawing diagrams.

3. A sketchpad at the lower left corner for gesturing general purpose editing commands such as Undo, Redo, as well as all menu commands by handwriting.

4. An InterViews standard *panner* at the lower right corner for zooming and scrolling.

6.4 Summary

Our prototype Handi implementation benefits greatly from both C++ and Unidraw. C++ allowed us to express the implementation directly in object-oriented terms, in particular, the excellent expressivity of inheritance and polymorphism makes the code more readable, maintainable, and extensible.

Unidraw provided a comprehensive object-oriented editor framework which eliminated considerable implementation work which is not directly in the scope of this research. As a result, the Handi library is much smaller than it would have been without using Unidraw. However, Unidraw does not support direct manipulation of *non-toplevel* graphical components. This made the main problem in the implementation of hierarchy related manipulations of subcomponents, in particular the graphics transformations.

The implementation of the recognizing subsystem was the hardest part of the development effort, although its responsible code is less than that for the editing subsystem. Many time was spent on the tuning of the low-level recognizer and the refinement of the gesture specification architecture. Because Unidraw supports basic connectivity semantics, the main effort was concentrated on the implementation of the hierarchy component.

The prototype demonstrates that the Handi architecture is realizable. In the next chapter, we describe two experimental applications which are used to prototype and to verify the Handi architecture and its implementation.

Chapter 7

Applications

Within the scope of this dissertation, we built two handsketch-based diagram editors in distinct work periods with different purposes. The statecharts editor was the first prototype editor with which we started our research [141], obtained many new ideas during its experimental development, and got deep understandings of gestural interfaces. The Petri net editor was developed after the Unidraw-based Handi implementation was available, and it benefited a lot from the data flow concept of Unidraw in its simulation and visualization. In a long iterative and incremental refinement process, we worked simultaneously on the Handi architecture and the two editors. Several ideas of the Handi architecture were gained in observing generalizable common attributes of these editors. This chapter describes the design and the implementation of these two handsketch-based diagram editors.

7.1　Statecharts Editor

7.1.1　The Statecharts Language

Statecharts have been introduced as a higraph-based visual formalism for specifying the behavior of complex reactive systems [44, 45]. They extend classical state transition diagrams by three essential elements which are hierarchy, concurrency, and communications. Statecharts are interesting because they were developed as a visual programming language not based on any textual language, and they have well-defined formal semantics [43]. They exploit the two-dimensional nature of pictures to express concurrency in a natural fashion, and present novel definitions of semantics in terms of shared variables, chain reactions, and simultaneous multiple messages. Statecharts have been used to specify complex avionics systems at the Israel Aircraft Industries [44], to generate hierarchical hardware controllers [48], and to design user interface management systems [132].

7.1.1.1 Graphical Syntax

In previous chapters, several examples of statecharts have been shown in figure 4.2 and 4.13. The syntax of statecharts is defined over the following basic sets of elements [43]: states, transitions, primitive events, primitive conditions and variables. Using these basic sets of elements, the extended sets of events, conditions, expressions, labels, and the interrelations connecting them, can be defined.

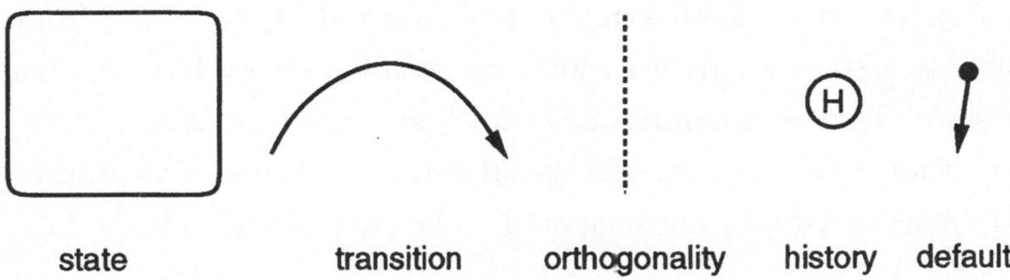

Figure 7.1: Graphical symbols used in statecharts

In order to build a handsketch-based statecharts editor, we concentrate on the graphical syntax of statecharts, and consider how the graphical components can be edited by using handsketches. Figure 7.1 depicts the graphical symbols used in statecharts. There are basically five[1] types of graphical symbols:

1. Rounded rectangles are used for representing states.

2. Arrow lines are used for representing transitions.

3. Dashed lines inside rounded rectangles are used to represent orthogonal states.

---

[1]There are several additional statecharts features which are not considered in this prototype. These additional features include the representations of condition, selection, timeout, and unclustering [44].

4. Circles labeled with the character 'H' represent the history symbols.

5. Black dots are used for indicating default states.

Based on these graphical symbols, statecharts define composition rules for building syntactical structures with well-defined semantics:

- The containment relationships between rectangles represent the hierarchy between corresponding states. Rectangles do not intersect each other. Each state has a name label at its upper left corner.

- Arrow lines can originate and terminate at any state and at any hierarchy level. An arrow line is labeled with an event description, optionally followed by a parenthesized condition. However, there are several exceptions: 1) Arrows from a history or a default symbol do not have transition labels. 2) An arrow which points to a default symbol is not allowed. 3) An arrow from a history symbol to a state is only used for a *continuation-arrow* from a default symbol to a state. There is maximal one arrow line of this type in each state. 4) There are no transitions between orthogonal states. 5) Arrows from and to the same state (loop transition) are suggested to be outside the state.

- Dashed lines are used to *separate* an xor state into several orthogonal states which are conceptually identical with xor states represented by rectangles.

- Each state has maximal one history symbol and one default symbol. A history symbol can carry a '*' mark for history with hierarchy iteration.

7.1.2   Gestural Interface

One of the most important efforts for building Handi-based applications is choosing, designing, and implementing a set of editing gestures. As discussed in chapter5, there are two basic types of gestures: language-*dependent* gestures and language-*independent* gestures.

Language-dependent gestures are usually gestures for creating diagram components. Our strategy for choosing such gestures is using directly the graphical symbols of the corresponding diagram elements. If a graphical symbol is not easy enough to draw, a simplified form can be used. For example, we use a simple circle as the gesture for creating a history symbol without writing the character H. Handi supports many language-independent and frequently used gestures such as clear, select, and delete gestures which can be directly used with desired gesture shapes. In appendix A, we illustrate a part of the designed gestures which are implemented in the current prototype. These screen dumps give a feeling about how the user can edit a statechart in paper-like style. At first sight, our handsketch-based statecharts editor differs from a conventional statecharts editor [45] in that there is no tool palette of command buttons. Although, the Handi architecture allows that the same command can be invoked by using both command buttons and gestures, the current implementation supports only gesture commands.

7.1.3 Implementation

The statecharts editor represents a typical Handi application with certain functionality and complexity. The current implementation offers:

- a complete gesture set for creating and manipulating statecharts,

- basic menu commands for storing and loading edited statecharts as well as creating PostScript and netlist representations of statecharts,

- undo and redo of editing commands,

- recognition of handwritten characters in the sketchpad for invoking menu commands,

- incremental syntax checking after each handsketch,

- structure preserving direct manipulations. For example, if a state is dragged, its substates and all related transitions are updated automatically for maintaining the statechart structure.

The Handi library provides a powerful support for the implementation of the handsketch-based statecharts editor. Many classes of the sketching subsystem and of the low-level recognizer of the Handi system can be used directly without any modifications. The implementation efforts are limited to design several new classes for implementing the language-dependent gestures and components as shown in table

7.1. These new classes are derived from corresponding base classes of Handi, they reuse the basic functionality, and build only the statecharts specific components. Table 7.2 presents a breakdown of the statecharts editor implementation in classes and lines of source code. Due to the many reusable functionality of the Handi library, the implementation efforts and the code size of our handsketch-based statecharts editor are very limited with regard to its functionality.

Table 7.1: Statecharts editor classes

Handi Class	Statecharts editor class
NodeGesture	CreateStateGesture
	CreateHistoryGesture
	CreateInitGesture
CreateEdgeGesture	CreateTransGesture
Gesture	AddOrthogonalGesture
HierarchyComp	StatechartComp
NodeComp	StateComp
	HistoryComp
	DefaultComp
EdgeComp	TransitionComp
HierarchyView	StatechartsView
NodeView	StateView
	HistoryView
	DefaultView
EdgeView	TransitionView
Command	AddOrthogonalCmd
DiaEditor	StatechartsEditor
DiaCreator	StatechartsCreator
NLHierarchy	NLStatecharts
NLNode	NLState
	NLDefault
	NLHistory
NLEdge	NLTransition

7.1.3.1 Gestures

In addition to the language-independent gestures such as `DeleteGesture`, `NameGesture` and `ClearGesture` which can be used directly to edit statecharts, the statecharts editor defines five subclasses of language-dependent gesture editing commands: `CreateStateGesture`, `CreateHistoryGesture`, `CreateDefaultGesture`, `CreateTransitionGesture`, and `AddOrthogonalGesture`. As discussed in chapter 4, a gesture class consists of three components: the gesture shape, the gesture constraints, and the gesture semantics. Currently, we use rectangle, circle, dot, opened, and line as the corresponding gesture shapes. The implementation of gesture constraints benefits by the high-level recognition functionalities of the Handi library for checking spatial relationships. The gesture semantics take the form of methods which create corresponding editing commands.

Because the gesture shape of `CreateTransitionGesture` overlaps the gesture shape of `AddOrthogonalGesture` (both are lines), their constraints checkers must distinguish between lines which connect two states and lines which insert orthogonal states. Further, the constraints checker of `CreateTransitionGesture` considers several quite sophisticated notations of arrow lines in statecharts. For example, arrow lines originating from a history symbol are not allowed *except* if an arrow points from a default symbol to a history symbol to form a continuation default notation (see figure A.2). And, there is no one transition which directs to a default symbol.

Table 7.2: Statecharts editor code breakdown

Module	Classes (number)	Interface (lines)	Implem. (lines)
Gestures	5	56	289
Component	10	171	471
Command	1	27	22
Editor	1	13	31
Creator	1	20	58
Netlist	5	74	173
Main	0	0	50
totals	23	360	1094

7.1.3.2 Components

The statecharts editor defines a new component class for each syntactical component of the statecharts language. Because the statecharts language is a typical HiNet diagram language, the Handi architecture supports enough functionality to simplify the component implementation. `StatechartComp` subclasses the `HierarchyComp` for implementing the statecharts hierarchy. `StateComp`, `History-Comp`, and `DefaultComp` are derived from the class `NodeComp` to realize their syntactical structures. `TransitionComp` subclasses `EdgeComp` to realize the statecharts transition syntax.

Each new component class has a corresponding view class. Because these view classes do not have essential extensions compared to their parent classes, their implementations are trivial. However, their existence is required by the general Unidraw architecture.

7.1.3.3 Commands

The `AddOrthogonalCmd` is the only new command class which is required for the statecharts editor. An object of this class is generated if a `CreateOrthogonalGesture` is recognized. The figures A.2 and A.5 illustrate several situations in which such a command is invoked by corresponding gestures. The add orthogonal command can transform an xor state into an AND state, or insert additional orthogonal substates into an existing AND state. Each add orthogonal command invokes several hierarchy manipulation operations which are implemented as methods of the classes `StatechartComp` and `StateComp`.

7.1.3.4 Editor, Creator, and Netlist

The class `StatechartEditor` subclasses `DiaEditor` for building its own gestural interface. In case that the default look and feel is used, the only effort to implement this class is to build the statecharts-dependent high-level recognizer. The following C++ code fragment illustrates the implementation of this class:

```
GestureSet* gesset = new GestureSet(this);

gesset->Add(new CreateTransGesture(StOpened));
gesset->Add(new CreateStateGesture(StQuadrilateral));
gesset->Add(new CreateHistoryGesture(StEllipse));
gesset->Add(new CreateInitGesture(StDot));
gesset->Add(new DeleteGesture(StComposite, MatchCross));
gesset->Add(new SelectGesture(StEllipse));
```

```
gesset->Add(new MoveGesture(StSharpArrow));
gesset->Add(new NameGesture(StLine));
gesset->Add(new AddOrthogonalGesture(StHorizontalLine));
gesset->Add(new AddOrthogonalGesture(StVerticalLine));
gesset->Add(new ClearGesture(StZForm\index{ZForm}, symbol_tree));
```

There is also a `StatechartsCreator`, derived from the Handi library's `Dia-Creator` class, that allows the statecharts editor to load an edited statechart diagram from a file.

Each statecharts component has a corresponding Netlist class for generating its netlist representation. The current implementation of these classes produces external presentation in the following EDIF-like format:

```
(STATECHART switch
      (STATE State1
            (INS   t2)
            (OUTS  t1))
      (STATE State2
            (INS   t1)
            (OUTS  t2))
      (TRANSITION t1
            (FROM  State1)
            (TO    State2))
      (TRANSITION t2
            (FROM  State2)
            (TO    State1)))
```

7.2   Petri Nets Editor

7.2.1   The Petri Nets Language

Petri nets [98] have been introduced by Carl Adam Petri as an extended graphical notation of finite automata for representing parallelism and data flow in 1962. They are easy understanding and precise, and come complete with an impressive body of research, accumulated over a period of thirty years. Petri nets might be considered as one of the best-known and best-understood visual languages with numerous applications, for example, system design with Petri nets [99], framework [97], and many others.

However, there are several drawbacks and limitations of the original Petri nets. As a result, various extensions and variants of the original place and transition nets have been introduced. Examples are predicated Petri nets [35], structured Petri nets [19], and hierarchical Petri nets [30]. Because the goal of this work is *not* the treatment of nets theories and semantics, the following discussions relate to only the graphical notations of Petri nets.

It should be mentioned that there are intricate hierarchy problems within hierarchical Petri nets which allow refinements of both places and transitions. A comprehensive discussion of these problems can be found in [30], whose details are not concerned to in the following considerations. Further, the graphical representation of hierarchy we use is *area-dominated*, that is, the spatial containment relationship is the main representation form of hierarchy. Similar to statecharts, the user is allowed to access places and transitions which belong to different hierarchical levels in the same window. This differs from approaches, for example [119], where the user can edit only one single Petri net diagram in one hierarchy level at a time. Editing commands such as "goto father", "goto son" are provided for *walking* through the diagram hierarchy.

Analyzing the graphics of these Petri net languages, it is obvious that the hierarchical Petri nets build the superset. The graphics of every specific Petri net language can be considered as hierarchical Petri nets with different constraints. For example, the structured Petri nets are specific hierarchical Petri nets in which only transitions can be refined into new nets. And an original Petri net is a hierarchical Petri net with a hierarchy depth of zero.

7.2.1.1 Graphical Syntax

Visual Alphabet For designing a handsketch-based Petri net editor, we concentrate on the graphical syntax of Petri nets, consider how the graphical components and the Petri net structure can be edited by using handsketches. Figure 7.2 shows the visual alphabet used in Petri nets. In the original Petri nets, small filled rectangles with constant size are used to represent transitions. After the structured and the hierarchical Petri nets have been introduced, unfilled rectangles with variable size are used to represented transitions and *macro transitions* [19]. Places are represented by circles with constant size in the original Petri nets and in the structured

Petri nets; and variable sized circles are used in hierarchical Petri nets in which places consist of Petri nets again. Small filled circles are used to represent tokens inside the corresponding places. Further, directed arrow lines are used to represent the connectivity between places and transitions.

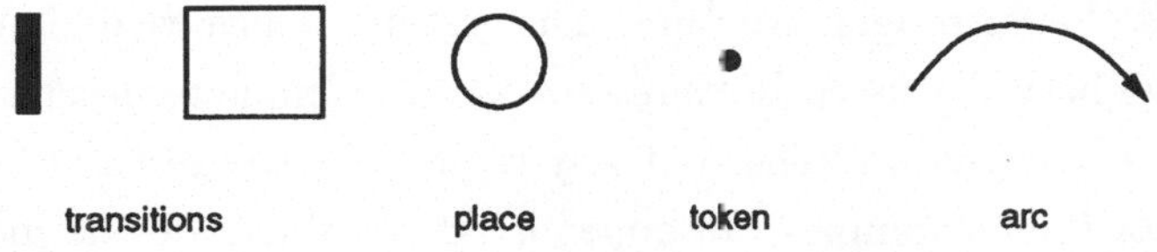

Figure 7.2: Graphical symbols used in Petri nets

Composition Rules Based on the visual alphabet of Petri nets, the graphical syntax rules can be defined. One of the reasons that Petri nets have been become so popular is that they have an easy understanding and simple graphical syntax which can be described as follows:

1. The graphical symbols of places and transitions do not intersect.

2. Both places and transitions may be associated with short name labels.

3. An arc connects exactly one place and one transition by using the coincidence relationship between its two endpoints and the boundaries of the corresponding place and transition. Arcs are only allowed from places to transitions and from transitions to places.

4. In hierarchical Petri nets, both places and transitions can build hierarchical structures by using the containment relationship between the related graphical symbols, that is, both transitions and places can be refined into complete new Petri nets.

5. Tokens are located inside circles of the corresponding places.

7.2.2   Gestural Interface

There are many tools, both graphical and non-graphical, developed for and based upon Petri nets, examples can be found in [125, 72, 66]. In this work, we concentrate on the graphical components of Petri nets and experiment on how handsketches can be used to edit the Petri nets structure. The gestures which we designed for editing Petri nets have been discussed in chapter 4 as an example to describe the concept of the high-level recognition. The best way to describe the gestural interface of the Petri net editor is by examples. In appendix B, we illustrate the most important gestures by a sequence of screen dumps in editing reasonable Petri nets.

As the statecharts editor presented in the previous section, our handsketch-based Petri net editor is a typical Handi application which consists of basic functionalities such as gestures for creating and manipulating Petri nets, menu commands for storing and loading Petri nets, PostScript and netlist generation, incremental syntax checking, structure preserving direct manipulations, as well as undo and redo.

Moreover, three additional features are experimented on this handsketched-based Petri net editor:

1. User customizations for different Petri nets variants with the same editor. For example, the user can start the Petri net editor by passing an option *-structured* to allow editing structured Petri nets, or by passing an option *-hierarchical* to allow editing hierarchical Petri nets.

2. Dynamical gestures such as adding tokens into places and firing enabled transitions. These dynamical gestures make the Petri net editor becoming an on-line Petri net simulator.

3. Data flow visualization by showing the token updates of corresponding places. Using colors and graphics to highlight enabled transitions. Places and transitions change their states both internal and external by using the object-oriented data flow abstraction designed in Unidraw.

7.2.3   Implementation

The Petri net editor creates several new classes shown in table 7.3. These classes are separated into two groups: the first group covers the *standard* classes of typical

Handi applications; the second group contains classes designed for implementing data flows of Petri nets. Class names in italic type denote classes provided by the Unidraw library, while the others in the first column are defined by the Handi library. Classes in the second column are derived new classes which reuse the Handi and Unidraw functionality to reduce the implementation efforts. Table 7.4 presents a breakdown of the Petri net editor implementation in modules and lines of source code.

Table 7.3: Petri net editor classes

Handi Class	Petri net editor class
NodeGesture	CreatePlaceGesture
	CreateTransGesture
CreateEdgeGesture	CreateArcGesture
Gesture	ActivateTransGesture
	AddTokenGesture
HierarchyComp	PetriNetComp
NodeComp	TransitionComp
	PlaceComp
EdgeComp	ArcComp
GraphicComps	Token
HierarchyView	PetriNetView
NodeView	TransitionView
	PlaceView
EdgeView	ArcView
GraphicViews	TokenView
Command	AddTokenCmd
	ActivateTransCmd
	TransmitPlaceCmd
DiaEditor	PetriNetEditor
DiaCreator	PetriNetCreator
NLHierarchy	NLPetriNet
NLNode	NLPlace
	NLTransition
NameVar	PlaceVar
	TransVar
TransferFunct	TFTransition
	TFPlace

Table 7.4: Petri net editor code breakdown

Module	Classes (number)	Interface (lines)	Implem. (lines)
Gestures	5	50	199
Component	10	174	423
Command	3	46	61
Editor	1	16	36
Creator	1	20	53
Netlist	3	54	160
Data flow	4	122	247
Main	0	0	53
totals	27	482	1232

7.2.3.1 Gestures

Additionally to standard language-independent gestures, the Petri net editor has several language-dependent gestures, each of which is implemented as a derived C++ class. `CreatePlaceGesture`, `CreateTransitionGesture`, and `CreateArcGesture` are gestures for creating Petri net components. `ActivateTransGesture` and `AddTokenGesture` can be used to simulate an edited Petri net.

As mentioned before, our Petri net editor allows user customizations for three different Petri nets variants (simple, structured, and hierarchical Petri nets). This is realized by using an editor parameter which controls the gesture semantics. When editing hierarchical Petri nets, the dimensions of handsketch-created places and transitions correspond to the size of the recognized handsketches. Hierarchical structures are recognized automatically by examing the containment relationships between the graphical symbols used in Petri nets. When editing simple Petri nets, both places and transitions have constant size. It is not possible to change it or to draw subcomponents inside such places and transitions. Because the only way to build hierarchy is to use the spatial containment relationship, the constraints on the object's size is used to define the Petri nets variants. In structured Petri nets, only transitions can build hierarchy, which is implemented with the same idea.

7.2.3.2 Components

Each symbol used in Petri nets defines a component and a view class to realize the
internal and external representation of the corresponding diagram component. For
implementing the Petri nets editor, each class defines several internal class members
to realize the data flow. For example, the class `TransitionComp` defines functions
such as `IsEnabled()` and `UseTokenFromJointedPlaces()`. The class `TokenRep` is
a graphical component class. Each place has an object of the class `TokenRep` to
display the correct number of small filled circles to represent the state of the place.

7.2.3.3 Commands

The Petri net editor defines three new commands. `AddTokenCmd` increments the
token number of the place to which an "add token" gesture relates. A command
object of the class `ActivateTransCmd` represents the gesture semantics of an "activate transition" gesture. The execution of such a command decrements the token
numbers of the connected input places and increments the token numbers of the
connected output places. Enabled transitions are visualized with highlighting colors
and thick lines. Because an arc gesture can change the state of the connected transition, a command of the class `TransmitPlaceCmd` is always executed after creating
a new arc to keep the consistency of the Petri net visualization such as a correct
highlighting of enabled transitions.

7.2.3.4 Editor, Creator, and Netlist

The class `PetriNetEditor` builds the gestural interface for handsketch-based Petri
nets editing. `PetriNetEditor` inherits the basic interactor's composition of `Dia-
Editor`, and creates only the desired gesture objects for the language-dependent
high-level recognizer as follows:

```
GestureSet* gess = new GestureSet(this);

gess->Add(new CreatePlaceGesture(StEllipse));
gess->Add(new CreateTransGesture(StQuadrilateral));
gess->Add(new CreateArcGesture(StOpened));
gess->Add(new DeleteGesture(StComposite, MatchCross));
gess->Add(new SelectGesture(StEllipse));
```

```
gess->Add(new ActivateTransGesture(StSharpArrow));
gess->Add(new MoveGesture(StSharpArrow));
gess->Add(new AddTokenGesture(StDot));
gess->Add(new NameGesture(StLine));
gess->Add(new ClearGesture(StZForm\index{ZForm}, symbolTree));
```

The **PetriNetCreator** makes Petri nets which are created by the editor, persistent, by allowing to read and to write a Petri net from and to a disk file. **NLPetriNet**, **NLPlace**, and **NLTransition** derive proper Handi's netlist classes for creating external representations of Petri nets. The following is an example of our EDIF-like Petri net representation.

```
(PETRINET PN
    (LIVING NO)
    (TRANSITION T_0
        (INS  P_1 P_0)
        (OUTS  P_2)
        (SUBNET ()))
    (PLACE P_0
        (TOKENNO. 0)
        (INS ())
        (OUTS  T_0)
        (SUBNET ()))
    (PLACE P_1
        (TOKENNO. 1)
        (INS ())
        (OUTS  T_0)
        (SUBNET ()))
    (PLACE P_2
        (TOKENNO. 1)
        (INS  T_0)
        (OUTS ())
        (SUBNET ())))
```

7.2.3.5 Data Flow

Unidraw provides a good abstraction for object-oriented data flow programming by supporting connector semantics with transfer functions of state variables. Within our handsketch-based Petri net editor, we have done some experiments on the integration of the data flow programming together with the visualization of the dynamical states of places and transitions.

When `PlaceComp` is instantiated, a place object creates an instance of a `PlaceVar` state variable and an instance of a `TokenRep` graphical component, and binds them to itself. The `PlaceVar` stores a name (being derived from `NameVar`), an identifier (name with a counter), the token number, and the limitation of tokens. The `TokenRep` represents the token number with small filled circles inside the place. A `TransitionComp` object binds a `TransVar` to itself to store a name, an identifier, and the state of this transition (passive, enabled).

`PlaceComp` also defines a `TFPlace` transfer function which implements the effect of the state variable of each place. For example, an "add token" gesture increments the token number of a place, the transfer function transmits this information to all connected transitions. `TransitionComp` defines a `TFTransition` transfer function which implements the switching rules of a transition.

7.3   Summary

The statecharts editor and the Petri net editor demonstrate how Handi facilitates the design and implementation of handsketch-based diagram editors. Each of these experimental applications reflects a significant reduction in development efforts by reusing the functionality of the Handi library.

Each application has features that set it apart from the other, thereby offering different perspectives of Handi's capabilities. The statecharts editor has the more complex gesture specifications; its hierarchy and orthogonal states related gesture recognition in particular shows how Handi's high-level recognition model can support sophisticated gesture recognition. The Petri net editor offers some features of a visual programming environment. Dynamical gestures together with the data flow visualization shows that the Handi architecture can be used not only for editing static diagram structures, but also for *running* diagrams.

Chapter 8

Evaluation

The previous chapters presented our basic concepts, the Handi architecture, its implementation, and two experimental applications. This chapter attempts to evaluate this work from two different perspectives: from the programmer's view in building handsketch-based diagram editors with Handi; and from the end-user's view in working with such editors.

8.1   Building Handi Applications

The Handi architecture defines object-oriented abstractions that simplify the construction of handsketch-based diagram editors. As a matter of fact, Handi greatly simplified the implementation of our two experimental editors. Though these editors are not polished systems, they demonstrate that Handi can help to build practical applications with certain functionality and complexity.

As mentioned before, building handsketch-based diagram editors has mainly two basic difficulties: one is the recognition of handsketches as well as the integration of a gesture recognizer into a graphical editor; the other is that general editor frameworks do not support structure editors directly. We designed a hierarchical and incremental low-level recognizer for recognizing handsketched graphical symbols, and a syntax-directed high-level recognizer for recognizing handsketches to editing commands. These key software components are encapsulated into *basic building blocks* and provided as reusable classes of the Handi library.

From the programmer's view, Handi may be evaluated on three levels: the effort required to build new diagram editors, the effort required to add and to modify gestures in an existing Handi application, and the extensibility of Handi.

8.1.1   New Applications

In chapter 7, we described the design and implementation of two experimental Handi-based applications, a statecharts editor and a Petri net editor. The statecharts editor gives us an opportunity for comparing development efforts required for building handsketch-based diagram editors with and without Handi, because the author had implemented a handsketch-based statecharts editor (SCE) [142] to gain first experiences with gestural interfaces before the Handi's existence.

SCE roughly provides the same editing functionality as the Handi-based version, but SCE is implemented on top of Motif [94] and Xlib [36]. Motif provides only very limited functionality for implementing diagram editors. Its numerous widgets and callbacks are useful for designing the main window or diverse dialog forms, but these are only a small part of a handsketch-based diagram editor. SCE calls Xlib directly for implementing drawing statecharts graphics and direct manipulations such as rubberbanding. Much of the time was spent on implementing functionality of gesture-based creation and manipulation of statecharts structures. The implementation of SCE took about three months and contains roughly 15,000 lines of source code in C.

After the Handi library was implemented, we rebuild this statecharts editor in about three weeks to get the first running prototype, which consists of only 1,300 lines source in C++. The difference in source code size and development time between these two prototypes is considerable, but in fairness the following caveats are necessary:

1. The implementation of Handi is built on top of Unidraw which provides comprehensive support for building object-oriented graphical editors, so that about 40 percent of SCE's code is replaced by Unidraw's functionality. This includes composing, drawing, direct manipulating, storing, and loading of graphical objects.

2. Handi is partly designed and abstracted from experiences by prototyping SCE, which implicates that the functionality required by a statecharts editor is well supported by Handi.

3. The recognizing subsystem of Handi directly supports the recognition of most frequently used graphical symbols, so that a Handi application usually need not deal with the low-level recognition. In contrast, about 30 percent of SCE's code dedicates to the implementation of recognition routines of handsketched statecharts graphics.

Despite these qualifications we can still conclude that Handi substantially reduces the development efforts for building handsketch-based diagram editors. This is confirmed by the development of the handsketch-based Petri net editor once more. It took about four weeks to design and implement it, and it has almost the same size of code. Much of the development efforts for the Petri net editor was spent on the implementation of the data flow visualization and the dynamical semantics of Petri nets. This functionality is not considered by the statecharts editor. Generally speaking, the implementation of the statecharts editor required more efforts than the implementation of the Petri net editor, because the visual syntax of statecharts is more complex than that of Petri nets. It is to infer that the development efforts for a Handi-based diagram editor are proportional to the complexity of the diagram syntax.

Since Handi provides comprehensive programming abstractions of handsketch-based diagram editors, the development efforts for each Handi-based editor are reduced to the implementation of the language-dependent gestures and diagram components. Due to the many encapsulatable common features among diagrams, the Handi architecture settled many important design issues and decisions in form of basic protocols and reusable classes. The required implementation can easily be done by considering what are nodes and edges of the diagram, and how they can be edited by using which gestures. The experiences with the statecharts editor and the Petri net editor show that the language-dependent classes build in average only one additional hierarchy layer of classes with tiny implementation efforts.

8.1.2 Adding and Changing Gestures

New gestures can be added to existing Handi applications easily. This is because Handi provides a powerful and convenient mechanism for specifying gestures. Indeed, several gestures used in the statecharts editor and in the Petri nets editor were inserted after the implementation of their first prototypes. As discussed in chapter

4, defining a new gesture can be done by specifying the gesture shape, the gesture constraints, and the gesture semantics.

A gesture shape may be chosen from graphical symbols supported by the low-level recognizer. The current Handi library provides a powerful low-level recognizer being able to recognize about thirty different graphical symbols which are usually sufficient for building diagram editors. However, if other graphical symbols have to be used for some reasons, corresponding recognition routines can be implemented by reusing Handi's basic recognition utilities such as the feature detectors.

The Handi architecture defines protocols for specifying gesture constraints and gesture semantics in form of methods of corresponding gesture classes. The implementation of a new gesture class often benefits by the functionality of predefined gesture classes. Moreover, Handi provides a set of utilities to support examinations of spatial relationships between graphical objects which are the frequently used compositions for building visual syntax.

A useful functionality of Handi, which was not discussed deeply so far, is the so-called *simple gestures* for predefined and language-independent editing commands such as undo, redo, save, load, quit, and others. These gestures do not have gesture constraints, that means, the corresponding constraints checkers return always `complete`. Because simple gestures do not have constraints and the semantics of such gestures are general editing commands, their definitions are reduced to the specification of a gesture shape and a command name. Therefore, it is possible to add such gestures via resources to an existing Handi application. The sketchpad which recognizes handwritten characters is designed for sketching such gestures as shown in figure A.8. The resource specification for the statecharts editor is defined as follows:

```
scdraw.ClearGesture: Z
scdraw.UndoGesture: U
scdraw.RedoGesture: R
scdraw.NewGesture: N
scdraw.OpenGesture: O
scdraw.SaveGesture: S
scdraw.SaveAsGesture: A
scdraw.PostScriptGesture: P
scdraw.NetListGesture: L
scdraw.QuitGesture: Q
```

8.1.3 Extensibility

Handi is an object-oriented system. One advantage of object-oriented systems is that adaptation and extension are easy because of the basic properties of the object-oriented programming.   All three subsystems can be extended by designing new classes in appropriate class hierarchies. As an example for the extensibility of other subsystems, we discuss how the low-level recognizer can be extended.

The class-hierarchy designed in section 3.3.1 is considered as a basic set of frequently used symbols. For most diagram languages, they are sufficient to build a gesture set. In cases that new symbols are required, new classes can be derived from existing classes as follows: One finds a class in the hierarchy which is most similar to the new class, but more general. The new class can then be defined as the subclass of this class. Further, it is possible to use the low-level recognizer as a framework where other recognition algorithms can be integrated. For example, several irregular gestures can be considered as subclasses of `opened` and `closed`.

8.2   Partial Utilization of Handi

The object-oriented design of Handi in three subsystems makes it possible to utilize individual subsystems by applications which are not developed completely on top of Handi. There are several reasonable partial utilizations of Handi:

1. Using the sketching subsystem as a front-end to build diverse gesture-based systems and pen-based applications. For example, the classes `Pen`, `Ink`, and `SketchingArea` can be used for developing algorithms for recognizing arbitrary gestures, or for building applications to support unconstrained freehand writing and drawing such as Freestyle [69] showed.

2. Using the powerful recognizing subsystem of Handi in pen-based computing environments. To the best of our knowledge, our low-level recognizer is the first incremental recognizer to allow successive and multiple-stroke sketches. It is planned to integrate the recognizing subsystem, at least the low-level recognizer, in a pen-based computer system.

3. In domains outside the conceptual design, for example, in editing large schematic diagrams or VLSI layouts, the conventional graphical user interface may

be more precise and appropriate than a handsketch-based user interface. The editing subsystem can be used to build diagram editors with conventional graphical user interfaces in the non-conceptual design domain. For example, our diagram abstractions can be used to build schematic diagram editors, because a schematic can be considered as nodes and edges as well.

Besides these possibilities, the author tried to integrate the freehand drawing functionality into the well known **idraw** editor [129] by using the sketching and the recognizing subsystems of Handi. This was done surprisingly in just twelve statements by adding a **Draw** and an **Edit** to the tool palette of **idraw**. And it works very well! Figure 8.1 illustrates the idraw editor with freehand sketching possibility.

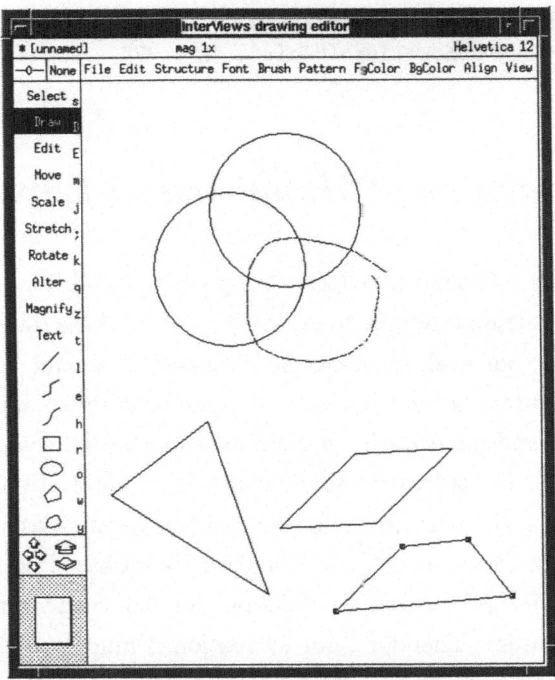

Figure 8.1: The freehand drawing functionality is integrated successfully in the well-established drawing editor **idraw**. This extended idraw allows the user create *beautified* pictures by using handsketches.

The following is the whole source code to integrate Handi's sketching and recognizing subsystems into `idraw`:

```cpp
#include <llr.h>
#include <pen.h>
...
SymbolTree* stree = new SymbolTree(_viewer);

GestureSet* gset  = new GestureSet(this);
gset->Add(new CreateCompGesture(StClosed));
gset->Add(new CreateCompGesture(StOpened));
Include(new Pen(new ControlInfo("Draw", "D", "D"), stree, gset), tools);

gset  = new GestureSet(this);
gset->Add(new DeleteGesture(StComposite, MatchCross));
gset->Add(new SelectGesture(StEllipse));
gset->Add(new MoveGesture(StSharpArrow));
Include(new Pen(new ControlInfo("Edit", "E", "E"), stree, gset), tools);
```

8.3   Performance of Handi-based Editors

The previous section discussed how Handi supports to build handsketch-based diagram editors. This section attempts to evaluate how well these Handi-based diagram editors work. Before our evaluation can be discussed, it should be mentioned that handsketch-based editors are not a mature domain with many applications so that a statistically valid and quantitative performance evaluation could be done by numerous comparisons. In contrast to some handwriting applications developed with pen-based computers, there are up to now only a few gesture-based drawing editors available as "toy" research prototypes. The two Handi-based editors presented in chapter 7 *may be* the first realistic applications for handsketch-based and syntax-directed diagram editors. Therefore, our evaluation is much subjective and reflects primarily the author's experiences with these Handi-based applications.

The performance of Handi-based diagram editors can be evaluated from several perspectives such as the recognition rate and the recognition speed, the tasks and actions a user performs to accomplish certain diagram editing operations, and the user's feeling of directness.

8.3.1   Gesture Recognition

Chapter 3 and chapter 4 presented our methods and algorithms for the low-level recognition and the high-level recognition, respectively. These two recognizers are the main components of every Handi-based application, their performances affect directly how well a Handi-based application satisfies a user. The experiences on working with the two experimental Handi applications show that both the low-level recognizer and the high-level recognizer produce satisfactory recognition results which can be considered in recognition rate and recognition speed.

8.3.1.1   Recognition Rate

The recognition rate of a gesture recognizer is the fraction of gestures that it correctly recognizes. Though Handi's recognizing subsystem consists of a low-level recognizer and a high-level recognizer, the recognition rate depends mainly on the low-level recognizer. This is because the correctness of a high-level recognizer should be guaranteed by the corresponding constraints checker, that is, each error produced by the high-level recognizer is considered as a program error, and not as a recognition error.

Usually, the recognition rate of a pattern recognizer is given as a percentage. While such numbers are very meaningful for character recognizers, in particular for off-line recognizers, they reflect only one aspect of an on-line handsketch recognizer. This is because the recognition rate of an on-line recognizer depends strongly on the accuracy of the drawing and on the input device. Measurements on the two Handi applications indicate that our low-level recognizer achieves a recognition rate of over 98%, if the user draws carefully. Further, a number of recognition errors can be corrected in a very easy way due to our *incremental* recognition method. A typical situation for this is, for example, the user wants to draw a circle to get a place in the Petri net editor. Occasionally, the stop point is not close enough to the start point, and the circle is not recognized. In this case, the user can draw additional strokes which complete such circles successively in the paper-like style.

8.3.1.2 Input Device and UNIX

It is worth to note that the evaluation is *not* done in a proper environment related to the input device (mouse) because pen-based computers are still not available for UNIX workstations on which the current applications are running. Although the author had investigated in combining a sonic digitizer as an input device with a graphical workstation in the early research period [143], these efforts were not continued due to the recent quick evolution of pen-based computers, and due to the hope to see pen-based UNIX workstations soon.

Some misclassifications of the low-level recognizer were largely beyond the control of the recognizer: there were problems using the mouse as a drawing device and problems using a user process in the non-real-time operating system UNIX to collect drawing data. The mouse of a Sun workstation has a tablet which is especially unsuitable for drawing, because the friction between the mouse and the tablet is sensible, and the mouse must be held orthogonal to the tablet for receiving correct feeling while drawing. Collection of move-events produces occasionally problems, if the workstation is overloaded by other processes. For example, the input handler usually loses characteristic point coordinates, if the user sketches too quick and the editor process does not get enough CPU.

8.3.1.3 Recognition Speed

The subjective impression indicates that the low-level recognizer is so fast that immediate feedback is always displayed after a stroke is finished. The user is not aware of any delay due to the low-level recognition. In comparison to the low-level recognizer, the high-level recognizer produces tiny delays in editing large diagrams. For getting some quantitative descriptions of the recognition speed, we put several output statements into the source code of corresponding methods to report the execution times both of the low-level recognizer and the high-level recognizer.

The quantitative measurements show that the recognition speed of the low-level recognizer depends on the size (number of digitized points) of each handsketch. The number of digitized point coordinates depends on the size of each sketch and the speed of sketching. The experiences of the author indicate that an average stroke consists of between three and forty point coordinates. Furthermore, the position in the symbol hierarchy (see page 49) of the input pattern determines how many decisions the low-level recognizer has to make to recognize it, which obviously influences the recognition speed.

Table 8.1 shows the average performance of the low-level recognition we measured on a Sun 4 (SparcStation 1, SunOS 4.1.1). We chose three kinds of symbols representing different positions in the symbol hierarchy. Dot is in the first layer of the symbol hierarchy, its recognition requires only one decision. Because the resolution of the used time measurement is one millisecond, and the output varies between zero and one millisecond, the recognition speed of a dot is less than one millisecond. Square is in the deepest layer, six decisions must be made to recognize a square. Circle is in the third layer of the symbol hierarchy, which represents the average number of required decisions.

Table 8.1: Recognition speed of the low-level recognizer

Stroke size (No. of points)	Dot	Circle	Square
		(milliseconds)	
≈ 2	< 1	-	-
≈ 20	-	4	7
≈ 50	-	7	20
≈ 200	-	12	85

Table 8.2 presents the measurements of the recognition speed of the high-level recognizer. Firstly, the recognition speed of the high-level recognizer depends on the number of existing diagram objects which the recognizer has to take into account. This is because the underlying constraints checker has to examine the spatial relationships between each gesture and all related graphical objects, which requires many time consuming calculations on graphical objects. Secondly, it depends on the complexity of the gesture constraints defined. Thirdly, the recognition speed of the high-level recognizer depends on the performance of the editor framework being used and the efficiency of the implementation language. The values in table 8.2 are measurements for several gestures defined for the statecharts editor. One important result which becomes obvious is that the recognition of connection gestures, such as "create transition" in the statecharts editor takes much more time than gestures such as "create state" due to the complexity of corresponding gesture constraints.

Table 8.2: Recognition speed of the high-level recognizer

Context No. of obj.	State	History	Orth.	Trans.
		(milliseconds)		
≈ 10	5	4	15	15
≈ 50	22	18	80	93
≈ 200	91	60	200	380

8.3.2 Human Factors

Human factors is a branch of applied psychology and a discipline to guide and enhance the process of designing user-centered interfaces. Though this work does not look in detail at this discipline, it helps us to evaluate the user interfaces of Handi-based diagram editors. Since a handsketch-based interface is a kind of direct manipulation interfaces, we use the criteria provided by Hutchins et al [51] to evaluate our interfaces.

8.3.2.1 Input Technique

Each input technique has its own strengths and weaknesses, just as each application has its own unique demands. The essential point on evaluating whether an input technique is a *proper* technique, is to check how well it matches the requirements of the application. The sketching input technique is particularly appropriate for editing diagrams because the main task in this domain is sketching. It is obvious that a *sketch*-based interface is closely matched to a *sketch*-dominated application.

As a matter of fact, the experimental applications are well accepted by the author's colleagues at Cadlab, and as a brilliant demonstration at the International Workshop on Advanced Visual Interfaces [144]. Experiences show that gestural interfaces attract both novel and expert users. One important advantage of this input technique is that the input is modeless, and thus the user draws in the same way as always. It is not required to know the current edit mode because only one sketching mode is available. The user can therefore concentrate on the real work.

8.3.2.2 Directness

One important criterion of a direct manipulation interface is the feeling of directness. Using gestural interfaces for editing diagrams, there is little need for the mental decomposition of tasks into multiple commands with complex syntax. On the contrary, each gesture produces a comprehensible result in the task domain that is immediately visible. The closeness of the task to the required action reduces the problem-solving load and stress. This makes reference to the concepts of *semantic directness* and *articulatory directness* suggested by Hutchins et al. in [51].

Directness has two separate and distinct aspects, the distance and the engagement [51]. While one involves a notion of the distance between the user's thoughts and the physical requirements of the system in terms of *gulf of execution* and *gulf of evaluation*, the engagement concerns the qualitative feeling that one is directly manipulating the objects of interest. Handi's user interface reduces the distance between a user's thoughts and a Handi-based diagram editor under use, because the translation from *what a user wants to do* to *what a user has to do* is simple and straightforward. This can be examined in the two forms of distance: semantic distance and articulatory distance.

Semantic Distance   The semantic distance deals with the relationship between the goals of a user and the meaning of his expression in an interface language, and the ease with which he can express his intentions in that language. Gestures in a diagram editor are handsketches which are exactly the same as the graphics of the underlying diagrams. The meaning of such *gestural expression* refers *directly* to what the user wants to say. The relationship between the user's intention and the organization of the instructions is clearly closer in a gesture-based interface than in a menu-based interface. For example, in the handsketch-based Petri net editor, a user's idea "create a place" can be expressed in form of drawing the place's circle at the desired position. The cognitive load to do this intention is minimal because the picture to be created exists in the same form in the user's brain due to the nature of visual languages. In contrast, within a conventional user interface, the user has to change the edit mode in which a place can be created, select an object which is the desired hierarchy parent, and then enter graphical attributes such as the position and dimension of the place's circle.

Articulatory Distance   The articulatory distance deals with the relationship between the meanings of expressions and their physical form, and the ease with which
a user's meanings can be translated into the form and appearance of both input and
output of a computer. Handi-based editors follow the principle of *what you draw
is what you get* (WY_DIWYG), which allows the user to create diagram elements in
exactly the same form as they are defined in terms of the visual alphabet, and as
they will look on the screen. Such interfaces accept as input the physical form of
drawing motions. This physical form of the input directly refers to the meaning of
expressions in a diagram editor. The desired operations are simply done by drawing
the appropriate gestures just like by using pen and paper. There are no command
syntax or command names to learn.

Direct Engagement   Laurel [71] indicated that a direct manipulation interface
should give the user a qualitative feeling that he is directly engaged with the control of the objects – not with the program, not with the computer, but with the
semantic objects of goals and intentions. This work follows this valuable suggestion
by providing incremental feedbacks. Both the low-level recognizer and the high-
level recognizer react upon immediately each user's sketch. The low-level recognizer
displays incrementally the recognition result after each stroke is finished, the user
sees immediately whether his sketch is recognized or not. A recognized gesture is
removed from the sketchpad, and the created diagram object is drawn immediately
with the correct attributes so that the user feels that he controls a discourse with
the diagram editor and manipulates the diagram elements themselves. For example,
in the Petri net editor, if the user connects a place and a transition, the state of
the transition updates automatically in terms of enabled or disabled. Therefore, the
Handi-based editor behaves in a way that the user can assume that the working
objects are things they refer to.

8.3.2.3 Mixed Interface

The choice of the "best" interaction style is and will remain a complex function of
the task, the user, the environment within which the user will work, and the tools
with which the user is to do the job [8, 111]. It is wrong to feel that the issue of style
has to do with making a choice between one interaction style versus another, for
example, taking one between gesture-based interface and menu-based interface. The

interface design idea of this work was to incorporate elements from several styles in a single interface, and the key issue lies on how to do this intelligently.

The integration of handsketch-based interface into the conventional graphical user interface supports a proper combination of gestural and other dialogue techniques. One of the reasons for this approach is that gestures are not always the best input techniques in any case and for any commands. The experiences with the experimental editors shows that for some specific tasks menu-oriented interface is rather superior than gestural interface. For example, to load an existing file, the user can select a file name with a "file selection box" easier than to write the file name by hand. Moving objects can be done better with rubberbanding than drawing an arrow gesture because precise movement can be made by using rubberbanding.

We have good experiences of proper combinations of gesture-based and menu-oriented interfaces. For example, we designed the sketchpad for invoking menu commands with gestures such as drawing a letter 'O' for opening a file. Although the command is invoked with a gesture, the file name can be still be selected by *click* menu-items.

8.3.2.4 Recall Syntax-Directed Editors

Syntax-directed editors designed for textual programming languages found only limited acceptance because the difficulties in use outweigh the benefits. Neal [86] presented an analysis based on a five-dimensional user model, the computer expertise, programming expertise, programming language expertise, learning strategy, and level of risk aversion.   In the following, we examined these aspects of our gesture-based and syntax-directed diagram editors.

Pen-based computers have just a pen as the single input device, thus the required computer expertise is obviously lower than that required to use keyboard and mouse with many buttons.   A significant advantage of visual languages is that they are easy to use and easy to learn. Therefore the required programming expertise and programming language expertise are reduced. Compared to a textual programming language which defines usually up to a hundred keywords, most diagram languages use only between three and ten graphical symbols.

Our incremental recognition strategy strongly supports learning by doing. On the one hand, the user can draw anything at any time just like using a pen to make sketches on a piece of paper. Inking and low-level recognition provide immediate feedback independent of the underlying diagram syntax. On the other hand, the user can learn quickly to draw syntactically correct gestures, because only such gestures can be recognized and transformed into diagram components. The experiences of several colleagues of the author show that both novice and experienced users accept the user interface favorably. Furthermore, our editor provides an "any time usable" undo gesture to encourage the confidence and control of the user.

Handsketch-based editing supports an interesting compromise between syntax-directed editing and general unconstrained editing. This is because, handsketch-based editing is not template-based, the user can draw everything what he wants, the editor is modeless. On the other hand, the automatic pattern recognition components of a handsketch-based diagram editor make an editor understanding the underlying visual syntax. Such a diagram editor is not a general purpose graphical drawing editor, because the handsketches are recognized incrementally as command gestures which create and manipulate the underlying diagrams. The point here is integration of pattern recognition techniques in human-computer interaction.

Chapter 9

Conclusion

9.1 Summary of Work

We began this work by identifying an important class of visual languages and a meaningful application of gestural interfaces for supporting such languages. We call this class of applications *handsketch-based diagram editors* which allow a user to create and manipulate pictures in these languages by handsketches just as using paper and pen. We analyzed the graphical syntax and the basic structures, and formally defined *HiNet* diagrams to abstract this class of visual languages.

Despite advances in pen-based computers and user interface technology, handsketch-based diagram editors are difficult to design and to build. We identified the main problems in constructing such editors and investigated proper solutions for each of the identified subproblems. One of the main problems is the gesture recognition which can further be separated into the low-level recognition and the high-level recognition. In order to satisfy the specific requirements of sketching diagrams, we developed an incremental and hierarchical low-level recognizer which transforms digitized point coordinates into regular graphical symbols. In the light of the common features with diagrams, we developed a high-level recognizer with a simple but powerful gesture specification mechanism and constraints-based recognition method.

Having a proper method for gesture recognition, we prompted our hypothesis that an architecture embodying the right abstractions would simplify developing handsketch-based diagram editors. The hypothesis in turn gave rise to experiments, in which we gained enough experiences with this class of applications, thus we formulated and specified the object-oriented Handi architecture, and implemented it. In designing the Handi architecture, we decided to make use of existing editor frameworks and built Handi on top of common programming abstractions for graphical user interfaces, which makes possible to integrate gestural interfaces into conventional graphical user interfaces without taking away any useful features of other user interface techniques.

In order to verify our reasoning, we built two experimental handsketch-based diagram editors, a statecharts editor and a Petri net editor. Handi greatly simplified their implementation. Though these editors are not polished systems, they demonstrate that Handi can be used to build practical applications with limited efforts and with amazing small size of code. The Handi library provides reusable functionality for recognizing handsketched graphical symbols, for specifying gesture commands, and for designing diagram components in form of predefined classes. The development efforts of a new handsketch-based diagram editor are reduced merely to implement the language-dependent functionality. Because Handi narrowed the design space for diagram editors by the basic programming abstraction such as NodeComponent, EdgeComponent, and HierarchyComponent, the implementation of the language-dependent functionality is a simple specialization of such base classes.

Handi-based applications are easy to use compared to diagram editors with conventional graphical user interfaces. The user does not need take care of the current editing mode like with a conventional graphical editor. Therefore, the user's concentration can be situated directly on thinking in the underlying diagram language. The user draws in free style, and a Handi-based editor enables what the user draws becomes what the user gets.

9.2   Open Problems and Future Work

This work presents a new application area of gestural interfaces which become just popular during the recent advances in pen-based computer technology. To the best of our knowledge, Handi is the first software architecture for building gesture-based diagram editors. Handi's concept and design deal with several novel problems due to its early birth and the relatively less investigated application domain. Handi provides a fertile ground for future research involving gestural interfaces and visual programming languages which become more and more attentions [61]. This section discusses several open problems and directions for future work.

Improving the Low-Level Recognition

The incremental and hierarchical low-level recognizer is one of the most robust and well-tested software components of the Handi library, it is probably suitable for commercial usages. However, its performance can be improved by considering some of the following remained problems.

1. In the current implementation of the hierarchical recognition, irreversible decisions are made at each hierarchy level by creating a more specific object and destroying the more general object. This has the drawback that early recognition errors lead the recognition process in wrong subhierarchy. For example, if a bad drawn circle is decided to be an opened symbol in the first hierarchy level because the distance between the two endpoints is big, this handsketch cannot be recognized because the symbol hierarchy under **opened** does not contain circle. This problem can be solved by storing a *recognition history* which maintains all decisions together with a *confidence value* to enable backtracking or "looking last decision" when needed.

2. Within our hierarchical recognition strategy, the implementation of each local classification function directly influences the recognition rate and recognition speed. Currently, the local decisions are mainly made in a hard coded sequence as shown in chapter 6. A reasonable improvement can be made by using well-established statistical recognition method in each local recognition routine. Moreover, in each local decision routines, many experimental thresholds are used, their values can be further improved with more experiments.

3. In this work, we had less experiments with character recognition because character recognition becomes to be system software of pen-based computers. It is obvious that a tight integration between graphics recognition and character recognition are necessary. It is thinkable to integrate a character recognition method in our graphical symbol recognition because characters are graphical symbols as well.

Limitation in Gesture Specification

One significant advantage of our incremental recognition is the elimination of explicit closures for multiple-stroke gestures, that improves the feeling of directness. This is greatly supported by the tight communications between the low-level recognizer and the high-level recognizer. However, the incremental recognition strategy requires that gestures must be *defined recognizable*. In other words, in specifying the gestures, one must consider that each of them can be distinguished from the other anyhow. For example, in case that both rectangles and rectangles with double lines are used as two different gestures with the same gesture constraints, the recognizing subsystem can only recognize the simple rectangle gesture, because the high-level recognizer transforms every complete rectangle gesture immediately into editing commands so that a double-lined rectangle can never be constructed.

This limitation in gesture specification is till now never become a real problem due to the following reasons: On the one hand, the structural information of the underlying diagram helps usually the recognition, for example, in the statecharts editor, lines are used successfully both for creating transitions and for creating orthogonal states. On the other hand, there are enough usable graphical symbols so that some equivalent symbol can still be used. In spite of that, this problem can be solved by allowing the recognizing subsystem to access the last created diagram element as a kind of stroke object. In this way, the last recognized diagram component can be transformed into another diagram element in the same way as that of symbols, in case that it is needed.

Gesture Description Language

Handi greatly supports the development of handsketch-based diagram editors. Though, the remained programming task is limited, the editor designer has to program the language-dependent functionality, for example the gesture constraints, which requires knowledge of some system details. A step beyond programming per hand is to follow the generator approach to generate this code automatically. To do this, we must have a language in which we can specifying the underlying diagram syntax, semantics, and the gesture editing commands. Our basic concept for specifying each gesture in its shape, constraints, and semantics provides a good ground for developing corresponding language elements. Szwillus [119] presented an excellent concept for generating graphical structure editors (GEGS) together with a powerful input language, in which our concept for gesture specification and gesture recognition may be integrated.

General Usability of Gestural Interfaces

This work showed that gestural interfaces can be used in sketching diagrams in conceptual design. However, whether handsketch-based interfaces generally improve task performance over conventional graphical user interfaces is a question that requires further research. Moreover, pen-based computers become just commercially available, but pen-based computers for UNIX are still not available. During this work, we didn't have any possibilities to test our Handi-based applications with a real pen-oriented environment. Therefore, an important further work is to gain experiences with pen-based computing environments. Because the current implementation runs only on UNIX and the X Window System [36], a first step may be done to use a pen-based computer as an X terminal and run Handi-based applications in a connected UNIX workstation.

Architecture Extensions

The current editing subsystem of Handi supports the programming abstraction of HiNet diagrams. It is worth to investigate whether other visual language, for example, pictorial janus [54] can also be supported by Handi. A number of extensions must be considered both in the external graphical representation and the internal structure representation. Considering the visual syntax, other spatial relationships such as touch, intersect, must be implemented. Considering the internal representation, structures outside connectivity and hierarchy must be supported as well.

Visualization

In prototyping the Petri net editor, we integrated some data flow simulation gestures such as adding tokens into places, activating transitions. This is an example, how visual programming and visualization of data flow can be integrated into a single gesture-based application. The author believes that visual programming will become more attractive, if a visual program is directly executable such as illustrated in [55] "Draw a picture and looks it running" instead of "Draw a picture, parse it, run the compiled program". This relates the term of *executable graphics* [70] as well.

Bibliography

[1] Hubert Achthaler. Unscharfe Lesehilfe, Schrifterkennung mit Fuzzy Logic. *C't Magazin für Computer Technik*, (5):212–221, 1992.

[2] J. Allgayer et al. XTRA: a natural-language access system to expert systems. *International Journal of Man-Machine Studies*, 31:161–195, 1989.

[3] Saul Altabet. The integration of multimedia and pen-based computers, panel session. In *Proceedings of the IEEE Workshop on Visual Languages*, page 171, 1991.

[4] J.C. Andershak, S. Lumelsky, I.F. Chang, T.P. Mears, A.A. Stone, and W.W. Stead. Medication charting via computer gesture recognition. IBM Research Report RC 16024(#69114), T.J. Watson Research Center, IBM Corporation, 1990.

[5] Farahangiz Arefi, Charles E. Hughes, and David A. Workman. Automatically generating visual syntax-directed editors. *Communications of ACM*, 33(3):349–360, 1990.

[6] James M. Avery. Interactive worksurface: An interface paradigm for sketchable things. Technical Report ACA-HI-127-88, MCC, May 1988.

[7] B. Backlund et al. Loggie – a language oriented generator of graphical interactive editors. Technical report, Swedisch Institute of Computer Science, Kista, August 1988.

[8] Ronald M. Baecker and William A.S. Buxton, editors. *Readings in Human-Computer Interaction: A Multidisciplinary Approach.* Morgan Kaufmann Publishers, Inc, 95 First Street, Los Altos, CA, 1987.

[9] G.D. Battista, A. Giammarco, G. Santucci, and R. Tamassia. The architecture of diagram server. In *Proceedings of the IEEE Workshop on Visual Languages*, pages 60–65. IEEE Computer Society Press, October 1990. Skokie, IL.

[10] A. Belaid and J.P. Haton. A syntactic approach for handwritten mathematical formula recognition. *IEEE Transactions on Pattern Analysis and Machine Intelligence*, 6(1), 1984.

[11] Grady Booch. *Object Oriented Design, with applications.* The Benjamin/Cummings Publishing Company, Inc. 1991.

[12] B.M. Brown, T.H. Fay, and C.L. Walker. Handprinted symbol recognition system. *Pattern Recognition*, 21:91–118, 1988.

[13] W. Buxton. An informal study of selection positioning tasks. In *Proceedings of Graphics Interface '82*, pages 323–328, 1982.

[14] W. Buxton. Lexical and pragmatic considerations of input structures. *Computer Graphics*, 17(1):31–37, 1983.

[15] W Buxton. Chunking and phrasing and the design of human-computer dialogues. *Information Processing 86*, pages 475–480, 1986.

[16] W. Buxton. There's more to interaction than meets the eye: Some issues in manual input. In D.A. Norman and S.W. Draper, editors, *User centered system design*, pages 319–337. Lawrence Erlbaum Associates, Hillsdale, NJ, 1986.

[17] Robert M. Carr. The point of the pen. *BYTE*, February 1991.

[18] Shi-Kuo Chang, Michael Tauber, Bing Yu, and Jing-Sheng Yu. A visual language compiler. *IEEE Transactions on Software Engineering*, 15(5), May 1989.

[19] L.A. Cherkasova and V.E. Kotov. Structured nets. *Mathematical Foundations of Computer Science, Lecture Notes in Computer Science*, (118):242–251, 1981.

[20] D. Chow and J. Kim. Paper like interface for educational applications. In *National Educational Computing Conference*, pages 337–344, Boston, Massachusetts, June 1989.

[21] Microsoft Corporation. *Microsoft Windows Kompendium, für das Betriebssystem Windows Version 3.1.* 1992.

[22] Gennaro Costagliola, Masaru Tomita, and Shi-Kuo Chang. A generalized parser for 2-d languages. In *Proceedings of the IEEE Workshop on Visual Languages*, pages 98–104, October 1991.

[23] C. Crimi, A. Guercio, G. Pacini, G. Tortora, and M. Tucci. Relation grammars for modeling multi-dimensional structures. In *Proceedings of the IEEE Workshop on Visual Languages*, pages 168–173, October 1990.

[24] Y. Le Cun. Handwritten digit recognition: Applications of neural network chips and automatic learning. *IEEE Communications Magazine*, pages 41–46, 1989.

[25] M.R. Davis and T.O. Ellis. The rand tablet: a man-machine graphical communication device. In *Proceedings of FJCC*, pages 325–331, 1964.

[26] Yannis A. Dimitriadis and Juan Lopez Coronado. A new mathematical editor, using an on-line symbol recognition. In *Proceedings of the Fourth International Conference on Human-Computer Interaction, Volume 2*, page 1321, Stuttgart, FRG, September 1991. Elsevier Science Publishers.

[27] B Duerr et al. A combination of statistical and syntactical pattern recognition applied to classification of unconstrained handwritten numerals. *Pattern Recognition*, 12:189–199, 1980.

[28] M.S. El-Wakil and A.A. Shoukry. On-line recognition of handwritten isolated arabic characters. *Pattern Recognition*, 22(2):97–105, 1989.

[29] Clarence A. Ellis and Simon J. Gibbs. Active objects: Realities and possibilities. In Won Kim and Frederick H. Lochovsky, editors, *Object-Oriented Concepts, Databases, and Applications*, chapter Future Research and Development in Object-Oriented Systems, pages 561–572. Addison-Wesley, 1989.

[30] Rainer Fehling. Hierarchische Petrinetze, Idee und grundlegende Struktur. Forschungsbericht 344, University of Dortmund, May 1990.

[31] F. Ferrucci, G. Pacini, G. Tortora, M. Tucci, and G. Vitiello. Efficient parsing of multidimensional structures. In *Proceedings of the IEEE Workshop on Visual Languages*, pages 105–110, 1991.

[32] H. Freeman. On the encoding of arbitrary geometric configurations. *IRE Trans. Electron. Comput.*, 10:260–268, 1961.

[33] T. Fujisaki, T.E. Chefalas, J. Kim, C.C. Tappert, and C.G. Wolf. Online run-on character recognizer: Design and performance. IBM Research Report RC 16322(#72451), T.J. Watson Research Center, IBM Corporation, 1990.

[34] J. Gall. *Systemantics: How Systems Really Work and How They Fail.* Ann Arbor, MI: The General Systemantics Press, 2nd edition, 1986.

[35] H.J. Genrich and K. Lautenbach. *System Modeling with High-level Petri Nets.* Theoretical Computer Science 13. Springer Verlag, 1981.

[36] J. Gettys, R. Newman, and R.W. Scheifler. *Xlib - C Language Interface.* MIT, X Window System, Version 11, Release 4, 1989.

[37] H. Goettler. Graph grammars and diagram editing. In *Proceedings of the Third International Workshop on Graph Grammars and their Applications to Computer Science*, pages 216–231. Lecture Notes in Computer Science, Springer, 1988.

[38] H. Goettler. Graph grammars, a new paradigm for implementing visual languages. In *Proceedings of EuroGraphics*, pages 505–516, 1989.

[39] E.J. Golin and S.P. Reiss. The specification of visual language syntax. *Journal of Visual Languages and Computing*, 1:141–157, 1990.

[40] Eric J. Golin. *A Method for the Specification and Parsing of Visual Languages.* PhD thesis, Brown University, 1991.

[41] Keith E. Gorlen, Sanford M. Orlow, and Perry S. Plexico. *Data Abstraction and Object-Oriented Programming in C++.* B.G. Teubner, Stuttgart and John Wiley & Sons, 1990.

[42] Klaus Gottheil, Hermann Kaufmann, Thomas Kern, and Rui Zhao. *X und Motif — Einführung in die Programmierung des Motif-Toolkits und des X-Window-Systems.* Springer Verlag, 1992.

[43] D. Harel. On the formal semantics of statecharts. In *Proceedings of the 2nd IEEE Symposium on Logic in Computer Science, Ithaca, New York*, 1987.

[44] D. Harel. Statecharts: A visual formalism for complex systems. *Science of Computer Programming*, 8:231–274, 1987.

[45] D. Harel et al. Statemate: A working environment for the development of complex reactive systems. In *Proceedings of the 10th IEEE International Conference on Software Engineering. Singapore*, April 1988.

[46] N. Heintze, J. Jaffar, S. Michaylov, P. Stuckey, and R. Yap. The clp(r) programmer's manual. Technical report, Department of Computer Science, Monash University, 1987.

[47] Richard Helm, Kim Marriott, and Martin Odersky. Building visual language parsers. In *Proceedings of the Conference on Human Factors in Computing Systems (CHI)*, pages 105–112, 1991.

[48] Peter Hennige. Generation of hierarchical controllers from specifications via statecharts. Technical Report Cadlab Report 8/90, Cadlab, 1990.

[49] J. Hollan et al. An introduction to HITS: Human interface tool suite. Technical Report ACA-HI-406-88, MCC, December 1988.

[50] Mamoru Hosaka and Fumihiko Kimura. An interactive geometrical design system with handwriting input. In B. Gilchrist, editor, *Proceedings of IFIP 1977*, pages 167–172. North-Holland, 1977.

[51] E.L. Hutchins, J.D. Hollan, and D.A. Norman. Direct manipulation interfaces. In D.A. Norman and S.W. Draper, editors, *User Centred System Design*, pages 87–124. Lawrence Erlbaum Associates, Publishers, Hillsdale, NJ, 1986.

[52] K. Iwata et al. Recognition system for three–view mechanical drawings. In *Pattern Recognition, 4th International Conference Proceedings, U.K.*, March 1988.

[53] J.C. Jackson and R.J. Roske-Hofstrand. Circling: A method of mouse-based selection without button presses. In *Proceedings of the Conference on Human Factors in Computing Systems (CHI)*, pages 161–166, 1989.

[54] Ken Kahn. Concurrent constraint programs to parse and animate pictures of concurrent constraint programs. Technical Report SSL-91-16, P91-00143, System Sciences Laboratory, Palo Alto Research Center, 1991.

[55] Kenneth M. Kahn, Vijay A. Saraswat, and Volker Haarslev. Pictorial janus: Eine vollständig visuelle Programmiersprache und ihre Umgebung. In *Jahrestagung der Gesellschaft für Informatik*, pages 427–436, 1991.

[56] H. Kangassalo. Concept D: A graphical language for conceptual modeling and data base use. In *Proceedings of the IEEE Workshop on Visual Languages*, pages 2–11, 1988.

[57] A. Kankaanpaa. FIDS—a flat-panel interactive display system. *IEEE Computer Graphics and Applications*, pages 71–82, March 1988.

[58] A. Kay and A. Goldberg. Personal dynamic media. *IEEE Computer*, 10(3):31–41, 1977.

[59] D.D. Kerrick and A.C. Bovik. Microprocessor–based recognition of hand-printed characters from a tablet input. *Pattern Recognition*, 21(5):525–537, 1988.

[60] J. Kim. Gesture recognition by feature analysis. IBM Research Report RC12472, T.J. Watson Research Center, IBM Corporation, January 1987.

[61] Takayuki Dan Kimura et al. Pen-based user interface, panel session. In *Proceedings of the IEEE Workshop on Visual Languages*, pages 168–173, 1991.

[62] Keiji Kobayashi, Fumio Yoda, et al. Recognition of handprinted kanji characters by the stroke matching method. *Pattern Recognition Letterns*, 1(5):481–488, July 1983.

[63] Haruhiko Kojima and Tohru Toida. On-line hand-drawn line-figure recognition and its application. In *9th IEEE International Conference on Pattern Recognition*, pages 1138–1142, November 1988.

[64] Keiji Kojima and Brad A. Myers. Parsing graphic function sequences. In *Proceedings of the IEEE Workshop on Visual Languages*, pages 111–117, 1991.

[65] Glenn E. Krasner and Stephen T. Pope. A description of the Model-View-Controller user interface paradigm in the Smalltalk-80 system. *Journal of Object Oriented Programming*, 1(3):26–49, August 1988.

[66] Elisabeth Kupitz and Jürgen Tacken. DECOR – tightly integrated design control and observation. Technical Report 17/92, Cadlab, 1992.

[67] Gordon Kurtenbach. Making marks self-revealing. *SIGCHI Bulletin*, (October), 1991.

[68] Gordon Kurtenbach and Bill Buxton. GEDIT: A test bed for editing by contiguous gestures. *SIGCHI Bulletin*, (April), 1991.

[69] Wang Human Factors Laboratory. Freestyle multimedia communication system. Demonstrations on the CHI'89 Conference on Human Factors in Computing Systems, May 1989.

[70] F. Lakin. Spatial parsing for visual languages. In Shi-Kuo Chang, Tadao Ichikawa, and Panos. A., editors, *Visual Languages*. Plenum Press, 233 Spring Street, NY, 1986.

[71] Brenda K. Laurel. Interface as mimesis. In D.A. Norman and S.W. Draper, editors, *User Centred System Design*, pages 67–86. Lawrence Erlbaum Associates, Publishers, Hillsdale, NJ, 1986.

[72] M. Leszak and H. Eggert. Petri-Netz Methoden und Werkzeuge; Hilfsmittel zur Entwurfsspezifikation und -validation von Rechensystemen. In *Informatik Fachberichte 197*. Springer Verlag, 1988.

[73] James S. Lipscomb. A trainable gesture recognizer. *Pattern Recognition*, 24(9):895–907, 1991.

[74] H.C. Liu and M.D. Srinath. Corner detection from chain-code. *Pattern Recognition*, 23(1/2):51–68, 1990.

[75] R. Makkuni. A gestural representation of the process of composing chinese temples. *IEEE Computer Graphics and Applications*, pages 45–61, December 1987.

[76] Gale L. Martin, John Lovgren, and James A. Pittman. Improving generalization in neural net learning. Technical Report ACA-HI-362-88, Microelectronics and Computer Technology Corporation, October 1988.

[77] Gale L. Martin and James A. Pittman. The value of a second hidden layer in improving generalization of neural net learning. Technical Report ACA-HI-396-88, Microelectronics and Computer Technology Corporation, December 1988.

[78] James Martin. *Recommended Diagramming Standards for Analysts and Programmers, A Basis for Automation.* Prentice-Hall, Inc., Englewood Cliffs, New Jersey, 1987.

[79] J. McCormack. *X Toolkit Library - C Language Interface.* MIT, X Window System, Version 11, Release 4, 1989.

[80] Bertrand Meyer. *Object-Oriented Software Construction.* Prentice Hall, New York, N.Y., 1988.

[81] P. Morrel-Samuels. Clarifying the distinction between lexical and gestural commands. *Internal Journal of Man-Machine Studies*, 32:581–590, 1990.

[82] H. Murase and T. Wakahara. Online hand-sketched figure recognition. *Pattern Recognition*, 19(2):147–160, 1986.

[83] Brad A. Myers. A new model for handling input. *ACM Transactions on Information Systems*, 8(3):289–320, July 1990.

[84] Brad A. Myers. Taxonomies of visual programming and program visualization. *Journal of Visual Languages and Computing*, 1(1):97–123, March 1990.

[85] I. Nassi and B. Shneiderman. Flowchart techniques for structured programming. *ACM SIGPLAN Notices*, 8(8):12–26, 1973.

[86] Lisa Rubin Neal. User modeling for syntax-directed editors. In *Human-Computer Interaction – INTERACT'87*, pages 131–134, 1987.

[87] F.J. Newbery. An interface description language for graph editors. In *Proceedings of the IEEE Workshop on Visual Languages*, pages 144–149, October 1988.

[88] W. Newman and R. Sproull. *Principles of Interactive Computer Graphics.* McGraw-Hill Co., New York, first edition, 1973.

[89] Heinrich Niemann. *Pattern Analysis.* Springer Series in Information Sciences. Springer-Verlag, 1981.

[90] Andy Novobilski. Penpoint programming, an operating system built of small, interchangeable components. *Object Magazine*, 2(3), September 1992.

[91] Kazumi Odaka, Toru Wakahara, and Isao Masuda. Stroke-order-independent on-line character recognition algorithm and its application. *Review of the Electrical Communications Laboratories*, 34(1):79–85, 1986.

[92] G. Odawara, M. Tomita, K. Hattori, O. Okuzawa, T. Hirata, and M. Ochiai. A human machine interface for silicon compilation. In *Proceedings of the 25th Design Automation Conference*, pages 115–120, 1988.

[93] A. Okazaki et al. Knowledge-controlled pattern recognition technique for hand-drawn logic symbols. In *Proceedings of the IEEE Workshop on Computer Architecture for Pattern Analysis and Image Database Management*, pages 524–531, 1985.

[94] Open Software Foundation, Eleven Cambridge Center. *OSF/Motif, Programmer's Guide - Toolkit*, 1991. Revision 1.1.

[95] Theo Pavlidis. *Algorithms for Graphics and Image Processing.* Springer-Verlag, 1982.

[96] J. Pittman. Recognizing handwritten text. In *Proceedings of the Conference on Human Factors in Computing Systems (CHI)*, pages 271–275. ACM Press, 1991.

[97] Franz J. Rammig and Bernd Steinmüller. Frameworks und entwurfsumgebungen. *Informatik Spektrum*, (Band 15, Heft 1):33–43, February 1992.

[98] Wolfgang Reisig. *Petri Nets, An Introduction.* EATCS, Monographs on Theoretical Computer Science. Springer-Verlag, 1985.

[99] Wolfgang Reisig. *Systementwurf mit Netzen.* Springer Compass. Springer-Verlag, 1985.

[100] Jim Rhyne, Doris Chow, and Michael Sacks. The handheld electronic notepad. In *Proceedings of the 1991 X Technical Conference*, 1991.

[101] J.R. Rhyne. Dialogue management for gestural interfaces. *Computer Graphics*, 21(2):137–142, April 1987. Workshop on User Interface Software.

[102] J.R. Rhyne and C.G. Wolf. Gestural interfaces for information processing applications. IBM Research Report RC-12179, T.J. Watson Research Center, IBM Corporation, 1986.

[103] A. Rosenfeld and J.S. Weszka. An improved method of angle detection on digital curves. *IEEE Transactions on Computers*, September 1975.

[104] D.T. Ross. Structured analysis (SA): A language for communicating ideas. *IEEE Transactions on Software Engineering*, 3(1):16–34, 1977.

[105] Dean Rubine. The automatic recognition of gestures. Technical Report CMU-CS-91-202, School of Computer Science, Carnegie Mellon University, December 1991.

[106] Dean Rubine. Integrating gesture recognition and direct manipulation. In *Proceedings of the 1991 USENIX Technical Conference*, 1991.

[107] Dean Rubine. Specifying gestures by example. *ACM SIGGRAPH'91, Computer Graphics*, 25(4), 1991.

[108] Dean Rubine. Combining gestures and direct manipulation. Formal Video Program on the Conference on Human Factors in Computing Systems, CHI'92, 1992. Carnegie Mellon University.

[109] SAC Operator's Manual Version 2.1, SAC 200 Watson Boulevard, Stratford, Connecticut 06497. *GP-7 Grafbar Digitizer Operator's Manual*, March 1989.

[110] Dagmar Schmauks and Norbert Reithinger. Generating multimodal output — conditions, advantages and problems. In *Proceedings of the International Conference on Computational Linguistics (Coling-88)*, pages 584–589, 1988.

[111] Ben Shneiderman. *Designing the User Interface: Strategies for Effective Human-Computer Interaction*. Addison-Wesley Publishing Company, 1987.

[112] Nan C. Shu. *Visual Programming*. Van Nostrand Reinhold, New York, 1988.

[113] John Silbert. Issues limiting the acceptance of user interfaces using gesture input and handwriting character recognition. In *Proceedings of the Conference on Human Factors in Computing Systems (CHI)*, pages 155–158, 1987. Panel summary.

[114] Philip Spiby. Reference manual of the EXPRESS language. Technical Report TC184/SC4/WG 5, ISO, April 1991.

[115] Kozo Sugiyama and Kazuo Misue. Visualizing structural information: hierarchical drawing of a compound digraphs. Research Report 86, International Institute for Advanced Study of Social Information Science, Fujitsu Limited, 1989.

[116] Ivan E. Sutherland. Sketchpad: A man-machine graphical communication system. In *Proceedings of Spring Joint Computer Conference*, pages 329–346, 1963.

[117] Gerd Szwillus. GEGS – a system for generating graphical editors. In *Human-Computer Interaction – INTERACT'87*, pages 135–141, 1987.

[118] Gerd Szwillus. Editing graphical structures. In *Proceedings of the Third International Conference on Human-Computer Interaction, Volume I, Boston, Massachusetts*, pages 587–594, September 1989.

[119] Gerd Szwillus. Specification of Graphical Structure Editors: The Graphical Editor Generator System GEGS. Habilitationsschrift, University of Dortmund, May 1989.

[120] Gerd Szwillus. Supporting graphical languages with structure editors. In *Proceedings of EuroGraphics*, pages 517–528, 1989.

[121] C.C. Tappert. Cursive script recognition by elastic matching. *IBM Journal of research and development*, 26(6):765–771, November 1982.

[122] C.C Tappert, C.Y Suen, and T. Wakahara. The state of the art in on-line handwriting recognition. *IEEE Transactions on Pattern Analysis and Machine Intelligence*, 12(8):787–808, August 1990.

[123] Joseph A. Tatman and Ross Shachter. Dynamic programming and influence diagrams. *IEEE Transactions on Systems, Man, and Cybernetics*, 20(2):365–379, 1990.

[124] T. Teitelbaum and T. Reps. The Cornell Program Synthesizer: A syntax-directed programming environment. *Communications of the ACM*, 24(9):563–573, 1981.

[125] Isolde Toussaint. Algorithmen zur Vergröberung und Verfeinerung von netzartigen Diagrammen. GMD-Studien 90, Gesellschaft für Mathematik und Datenverarbeitung, October 1984.

[126] Frank Vahid, Sanjiv Narayan, and Daniel D. Gajski. SpecCharts: A language for system level synthesis. In *Proceedings of the CHDL 1991*, pages 145–154, 1991.

[127] Verfügbare Pen-Computer. *COMPUTERWOCHE*, 35(28. August), 1992.

[128] John M. Vlissides. Generalized graphical object editing. Technical Report CSL-TR-90-427, Computer Systems Laboratory, Standford University, June 1990.

[129] John M. Vlissides and Mark A. Linton. Unidraw: A framework for building domain-specific graphical editors. *ACM Transactions on Information Systems*, 8(3):237–268, July 1990.

[130] J.R. Ward and M.J. Phillips. Digitizer technology: Performance characteristics and the effects on the user interface. *IEEE Computer Graphics and Applications*, pages 31–44, April 1987.

[131] André Weinand, Erich Gamma, and Rudolf Marty. Design and implementation of ET++, a seamless object-oriented application framework. *Structured Programming*, 10(2):63–87, 1989.

[132] P.D. Wellner. Statemaster: A uims based on statecharts for prototyping and target implementation. In *Proceedings of the Conference on Human Factors in Computing Systems (CHI)*, pages 177–182, 1989.

[133] Tom Williams. Fuzzy logic simplifies complex control problems. *COMPUTER DESIGN*, pages 90–102, March 1991.

[134] Kent Wittenburg, Louis Weitzman, and Jim Talley. Unification-based grammars and tabular parsing for graphical languages. Technical Report ACT-OODS-208-91, MCC, June 1991.

[135] C.G. Wolf. Can people use gesture commands? *SIGCHI Bulletin*, 18(2):73–74, October 1986.

[136] C.G. Wolf. A comparative study of gestural and keyboard interfaces. IBM Research Report RC 13906(#62477), T.J. Watson Research Center, IBM Corporation, 1988.

[137] C.G. Wolf, J. Rhyne, L.A. Zorman, and H.L. Ossher. WE-MET (Window Environment-meeting Enhancement Tools). In *Proceedings of the Conference on Human Factors in Computing Systems (CHI)*, pages 441–442, 1991.

[138] C.G. Wolf and J.R. Rhyne. A taxonomic approach to understanding direct manipulation. IBM Research Report RC 13104, T.J. Watson Research Center, IBM Corporation, 1987.

[139] C.G. Wolf, J.R. Rhyne, and H.A. Ellozy. The paper-like interface. In *Proceedings of the Third International Conference on Human-Computer Interaction*, volume II, pages 495–501, September 1989.

[140] Yao Yong. Handprinted chinese character recognition via neural networks. *Pattern Recognition Letters*, 7(1):19–25, 1988.

[141] Rui Zhao. Paper-like interface for statecharts with graphical workstations. Technical Report 10/90, Cadlab, Joint Venture University of Paderborn - Siemens Nixdorf Informationssysteme AG, 1990.

[142] Rui Zhao. Using gestures for diagram editing. Technical Report 12/90, Cadlab, Joint Venture University of Paderborn - Siemens Nixdorf Informationssysteme AG, 1990.

[143] Rui Zhao. Paper-like interface for graphical workstations. In *Proceedings of the Fourth International Conference on Human-Computer Interaction, Volume 2*, page 1340, Stuttgart, FRG, September 1991. Elsevier Science Publishers.

[144] Rui Zhao. Gestural interfaces for diagram editors. In T. Catarci, M.F. Costabile, and S. Levialdi, editors, *Proceedings of the international Workshop on*

Advanced Visual Interfaces, pages 413–414, Roma, Italy, May 1992. World Scientific Series in Computer Science, Vol. 36.

[145] Rui Zhao. Incremental recognition of hand-sketched diagram graphics in gestural interfaces. In Russell Beale and Janet Finlay, editors, *Neural Networks and Pattern Recognition in Human Computer Interaction (CHI'91 Workshop)*, pages 143–162. Ellis Horwood, 1992.

[146] Rui Zhao. On-line geometry recognition using C++, an object-oriented approach. In *Proceedings of the 7th International Conference & Exhibition of Technology of Object-Oriented Languages and Systems, TOOLS 7*, pages 371–378, April 1992.

[147] Rui Zhao. Handi: A framework for building handsketch-based diagram editors. In *Proceedings of the 5th International Conference on Human-Computer Interaction*, Orlando, August 1993.

[148] Rui Zhao. Incremental recognition in gesture-based and syntax-directed diagram editors. In *Proceedings of the ACM Conference on Human Factors in Computing Systems (InterCHI'93)*, Amsterdam, 1993.

Appendix A

Statecharts Editing Scenarios

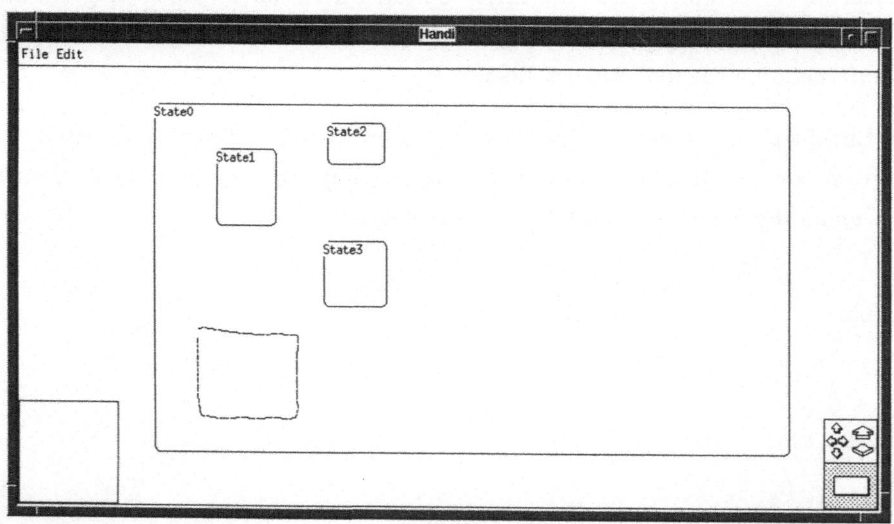

Figure A.1: Draw a rectangle to create a state. After this gesture is recognized, a state is displayed immediately. Hierarchy of states is recognized by using the containment relationships between states. This gesture will be recognized as a "create state" command and the new state will be inserted as a substate of State0.

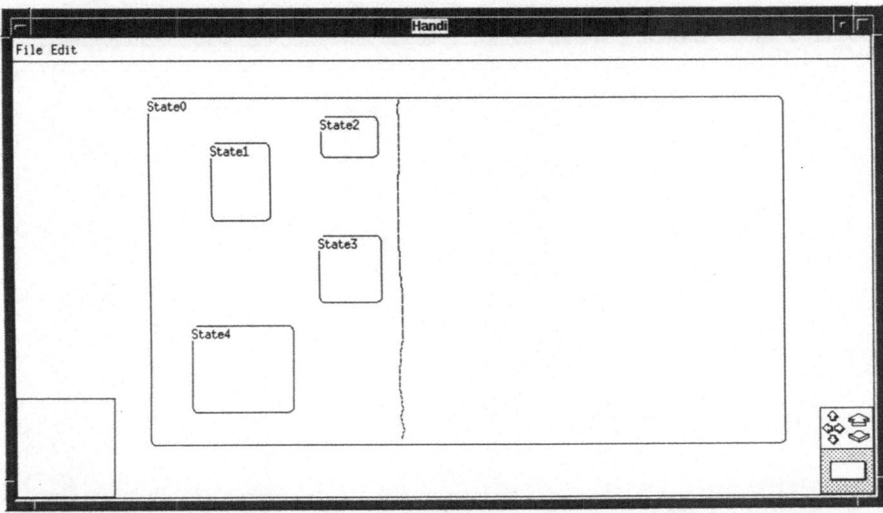

Figure A.2: Draw a straight line to create orthogonal states as substates of an xor state. The existing substates are reparented automatically.

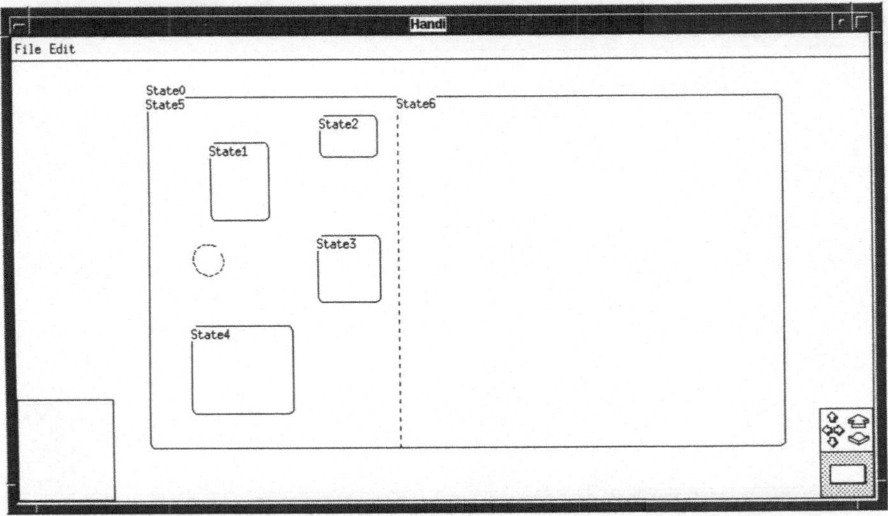

Figure A.3: Draw a circle inside a state to create a history symbol. The history symbol will be inserted in the smallest state enclosing this circle.

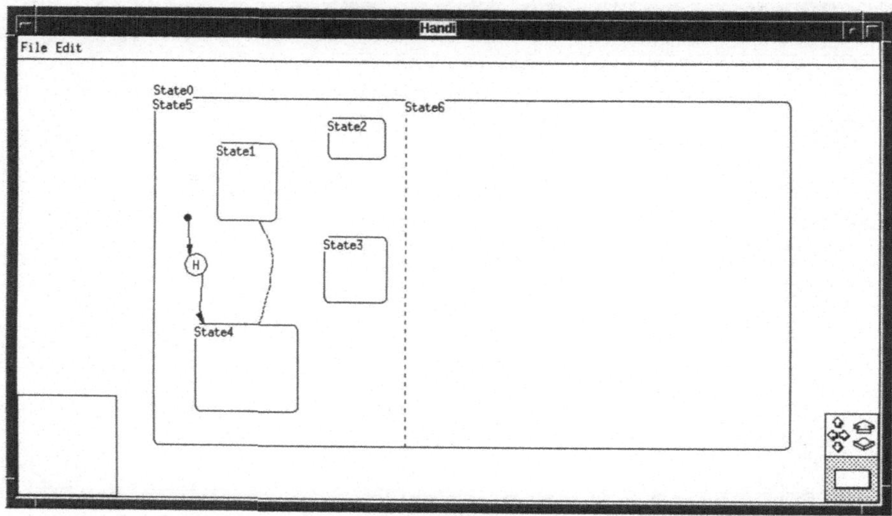

Figure A.4: Draw a connection line to create a transition. The drawing direction is recognized as the direction of the transition.

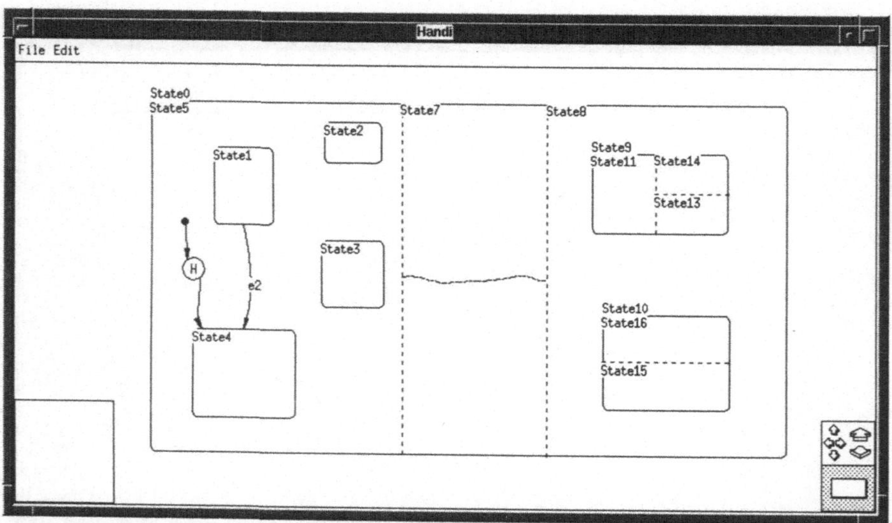

Figure A.5: The statecharts editor supports all possibilities to insert orthogonal states.

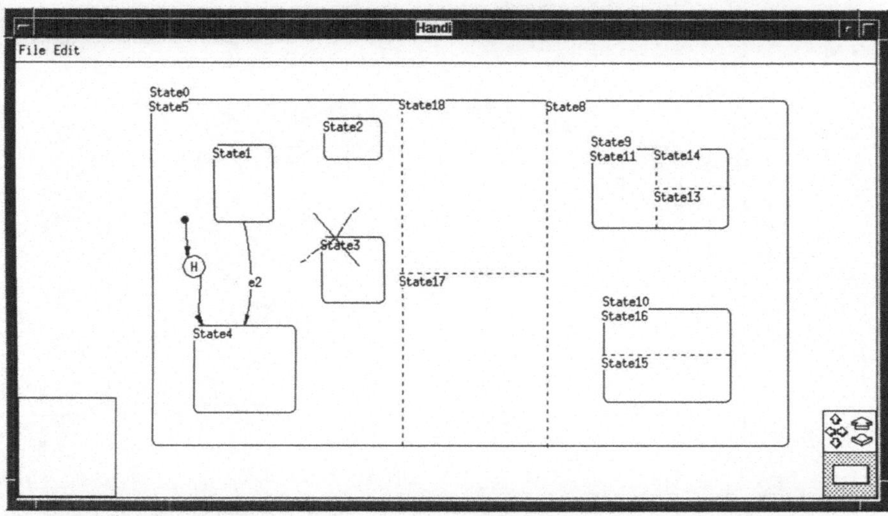

Figure A.6: Draw a cross symbol over an object to delete this object. The user does not need select it before.

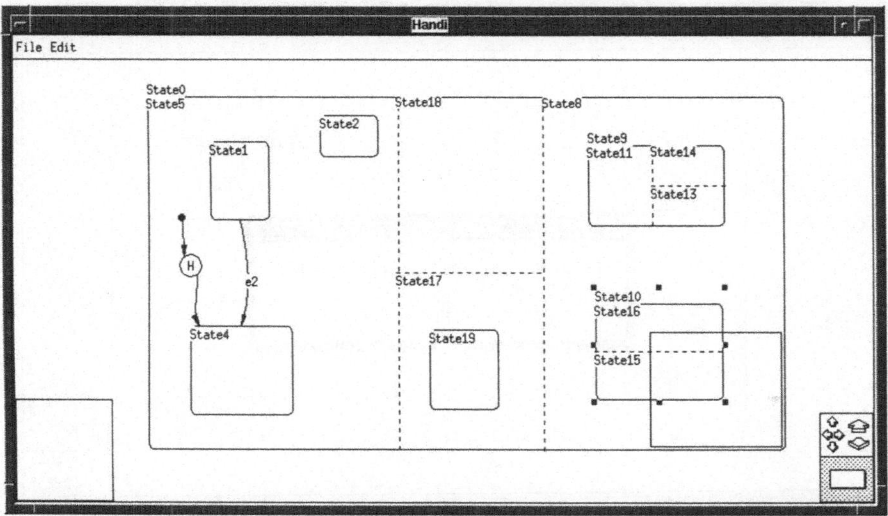

Figure A.7: Handi supports *syntax-constrained* direct manipulations such as states cannot be dragged outside the boundary of its parent state.

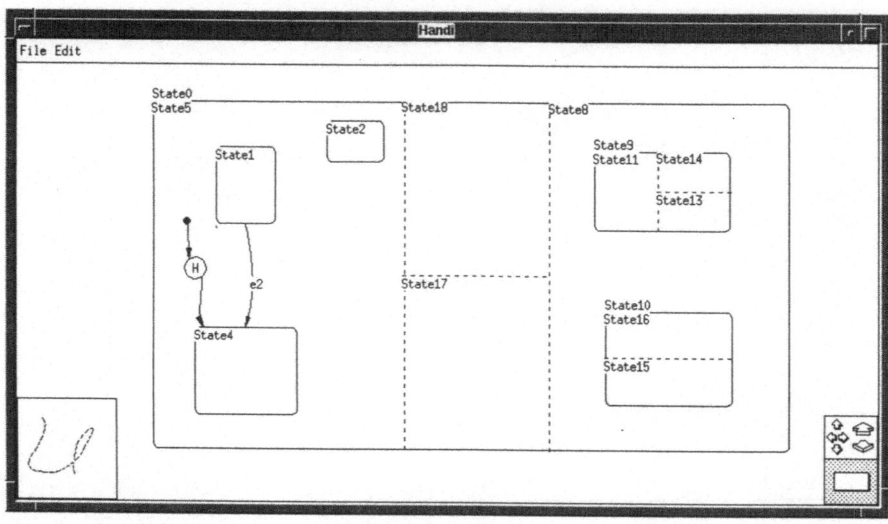

Figure A.8: Undo and all other File and Edit menu commands can be invoked by writing user defined *gesture-letters* in the sketchpad at the lower left corner.

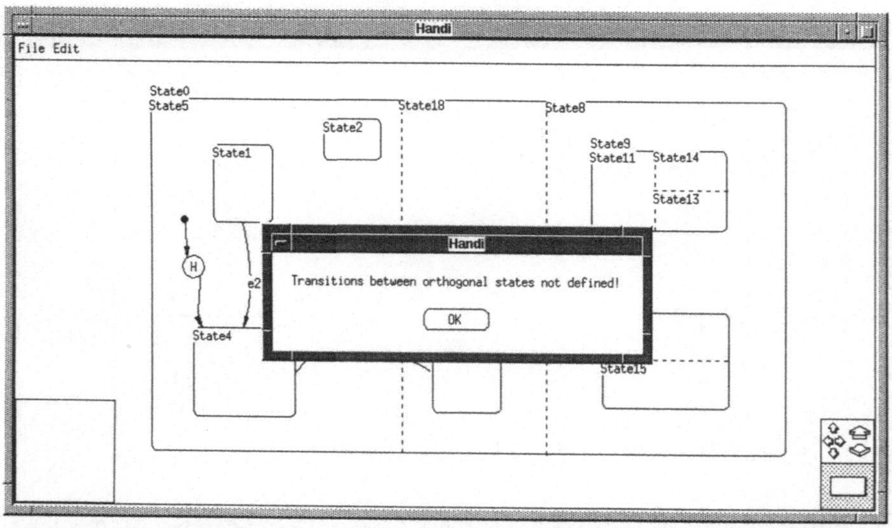

Figure A.9: Semantic errors such as transitions between orthogonal states are detected and indicated by corresponding error messages.

Appendix B

Petri Nets Editing Scenarios

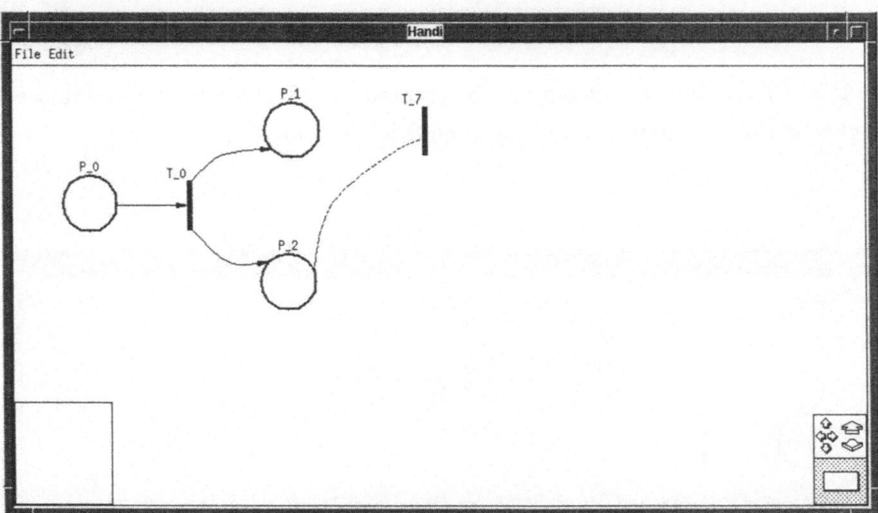

Figure B.1: Draw a freehand line to connect a place to a transition, or a transition to a place. The connection semantics are checked by the high-level recognizer. Wrong connection-gestures are rejected with corresponding error messages.

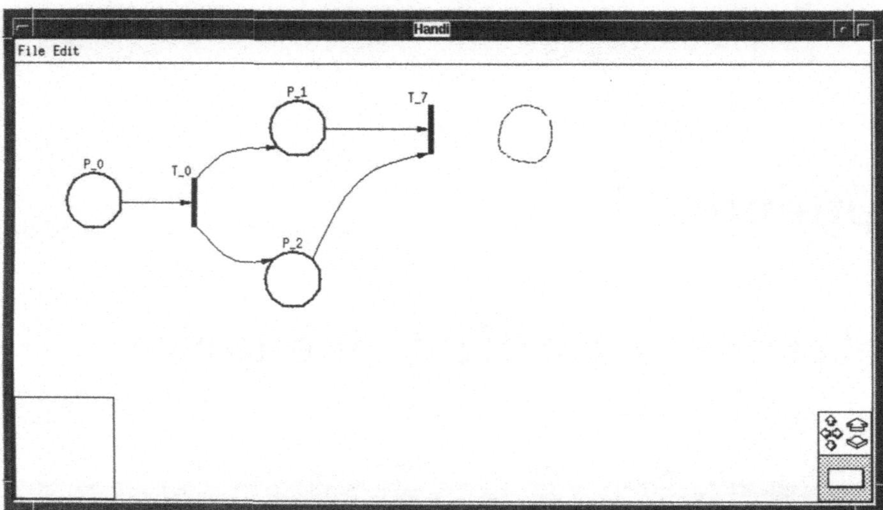

Figure B.2: Places and transitions can be created by sketching freehand ellipses and rectangles at the positions where they should be positioned.

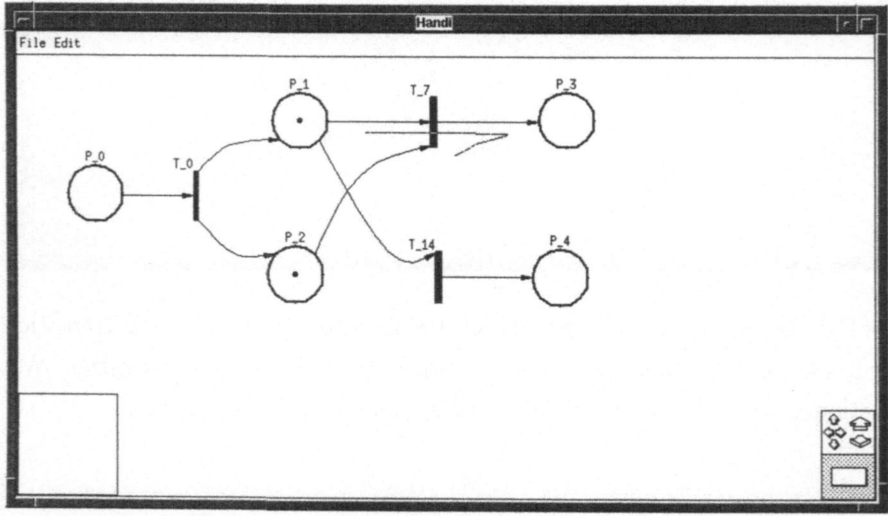

Figure B.3: The user can draw a dot to add a token into a place, and sketch an arrow to fire an enabled transition which is highlighted with thick line or with colors.

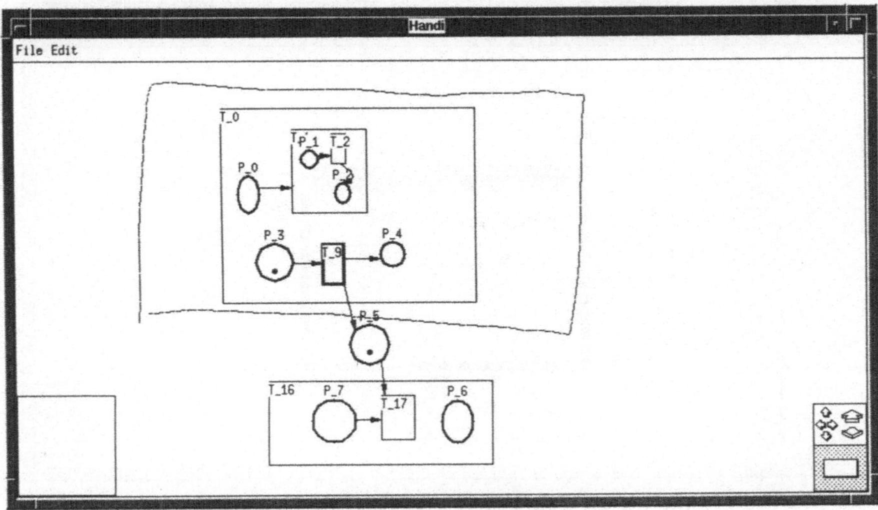

Figure B.4: This handsketched rectangle is recognized as a structured transition which increments the hierarchy level by one.

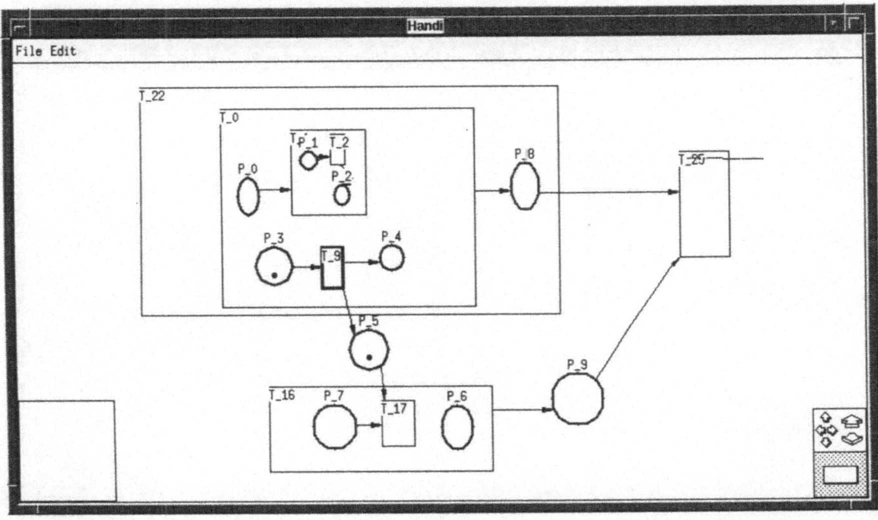

Figure B.5: Both places and transitions have default names which can be edited by using the name gesture, a line which intersects the corresponding text.

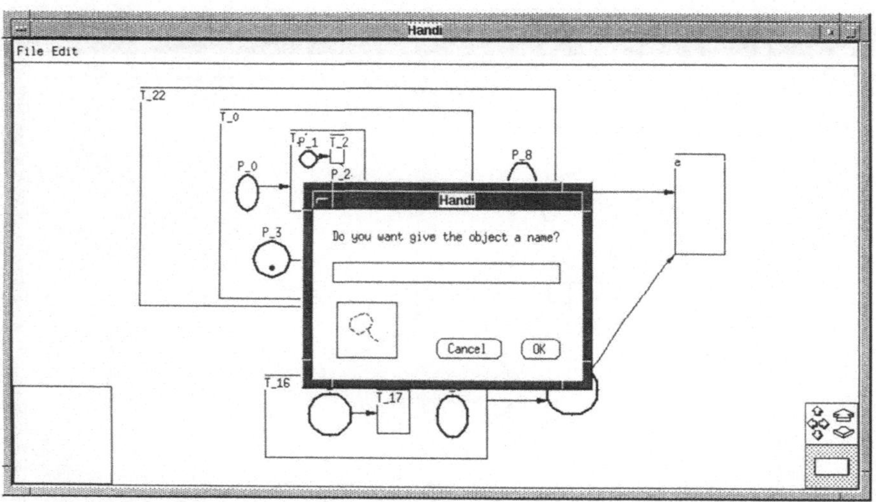

Figure B.6: After the name gesture is recognized, a dialog box is displayed with an integrated sketchpad which recognizes handwritten English characters.

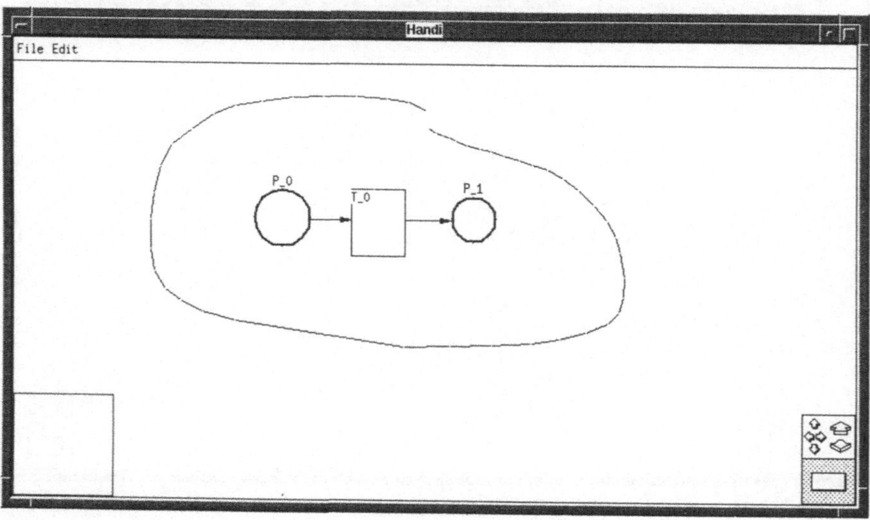

Figure B.7: Sketch a big circle to build a hierarchical place.

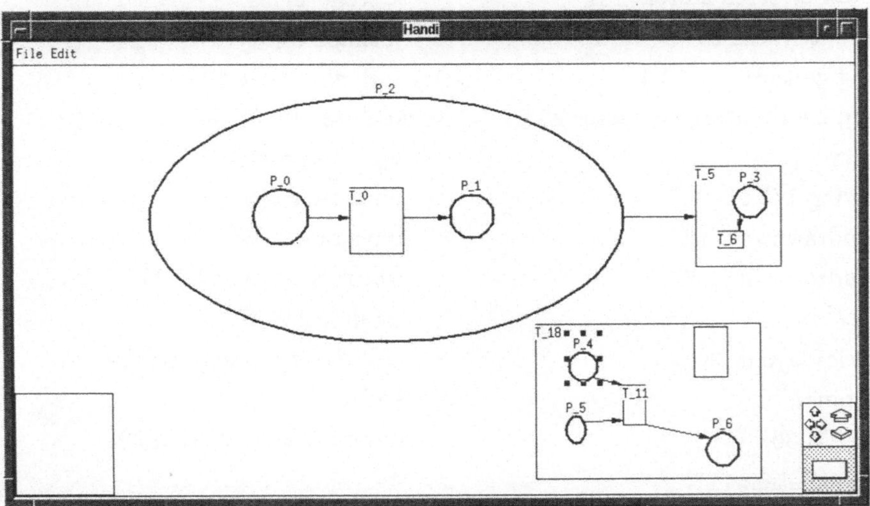

Figure B.8: Within Handi applications, the diagram structures such as hierarchy and connectivity constrain the dragging and stretching of diagram elements. Here, the place P_4 cannot be dragged outside the boundary of its parent-transition T_18.

Index